Federal Standards and Procedures for the National Watershed Boundary Dataset (WBD)

By the U.S. Geological Survey and the U.S. Department of Agriculture,
Natural Resources Conservation Service

Chapter 3 of

Section A, Federal Standards

Book 11, Collection and Delineation of Spatial Data

Techniques and Methods 11–A3
Third edition, 2012

U.S. Department of the Interior
U.S. Geological Survey

U.S. Department of the Interior
KEN SALAZAR, Secretary

U.S. Geological Survey
Marcia K. McNutt, Director

U.S. Geological Survey, Reston, Virginia: 2012
First edition: 2009
Second edition: 2011
Third edition: 2012

For more information on the USGS—the Federal source for science about the Earth, its natural and living resources, natural hazards, and the environment, visit http://www.usgs.gov or call 1–888–ASK–USGS.

For an overview of USGS information products, including maps, imagery, and publications, visit http://www.usgs.gov/pubprod

To order this and other USGS information products, visit http://store.usgs.gov

Suggested citation:

U.S. Geological Survey and U.S. Department of Agriculture, Natural Resources Conservation Service, 2012, Federal Standards and Procedures for the National Watershed Boundary Dataset (WBD) (3d ed.): U.S. Geological Survey Techniques and Methods 11–A3, 63 p.
Available on the World Wide Web at http://pubs.usgs.gov/tm/tm11a3/.

Acknowledgments

In 2009, the first edition of the "Federal Guidelines, Standards, and Procedures for the National Watershed Boundary Dataset" was produced and released through contributions by the Watershed Boundary Dataset National Technical Coordinators (WBD–NTC), as requested by the Subcommittee on Spatial Water Data (a subcommittee of the Federal Geographic Data Committee, Advisory Committee on Water Information), to update and clarify a previous guidance docu¬ment, "Federal Standard for Delineation of Hydrologic Unit Boundaries, Version 2.0, October 1, 2004." The second edition of the "Federal Standards and Procedures for the National Watershed Boundary Dataset (WBD)" was published in 2011. The WBD–NTC Team has first-hand knowledge of the use of both documents through their work with WBD creators from different agencies and programs and through regional workshops that contributed to coordinated watershed management. As in previous versions, this third edition of the "Federal Standards and Procedures for the National Watershed Boundary Dataset (WBD)," would not have been possible without the work of the U.S. Department of Agriculture, Natural Resources Conservation Service (NRCS), the WBD–NTC, and numerous prior contributors and reviewers.

Contents

Figures

Conversion Factors

Inch-pound to SI

Multiply	By	To obtain
Length		
foot (ft)	0.3048	meter (m)
mile (mi)	1.609	kilometer (km)
mile, nautical (nmi)	1.852	kilometer (km)
Area		
acre	4,047	square meter (m^2)
acre	.4047	hectare (ha)
acre	.4047	square hectometer (hm^2)
acre	.004047	square kilometer (km^2)
square mile (mi^2)	2.590	square kilometer (km^2)

SI to inch-pound

Multiply	By	To obtain
Length		
meter (m)	3.281	foot (ft)
kilometer (km)	.6214	mile (mi)
kilometer (km)	.5400	mile, nautical (nmi)
Area		
square meter (m^2)	10.76	square foot (ft^2)
square meter (m^2)	.0002471	acre

Federal Standards and Procedures for the National Watershed Boundary Dataset (WBD)

By U.S. Geological Survey and U.S. Department of Agriculture, Natural Resources Conservation Service

Abstract

The Watershed Boundary Dataset (WBD) is a comprehensive aggregated collection of hydrologic unit data consistent with the national criteria for delineation and resolution. This document establishes Federal standards and procedures for creating the WBD as seamless and hierarchical hydrologic unit data, based on topographic and hydrologic features at a 1:24,000 scale in the United States, except for Alaska at 1:63,360 scale, and 1:25,000 scale in the Caribbean. The data within the WBD have been reviewed for certification through the 12-digit hydrologic unit for compliance with the criteria outlined in this document. Any edits to certified data will be reviewed against this standard prior to inclusion. Although not required as part of the framework WBD, the guidelines contain details for compiling and delineating the boundaries of two additional levels, the 14- and 16-digit hydrologic units, as well as the use of higher resolution base information to improve delineations. The guidelines presented herein are designed to enable local, regional, and national partners to delineate hydrologic units consistently and accurately. Such consistency improves watershed management through efficient sharing of information and resources and by ensuring that digital geographic data are usable with other related Geographic Information System (GIS) data.

Terminology, definitions, and procedural information are provided to ensure uniformity in hydrologic unit boundaries, names, and numerical codes. Detailed standards and specifications for data are included. The document also includes discussion of objectives, communications required for revising the data resolution in the United States and the Caribbean, as well as final review and data-quality criteria. Instances of unusual landforms or artificial features that affect the hydrologic units are described with metadata standards. Up-to-date information and availability of the hydrologic units are listed at *http://www.ncgc.nrcs.usda.gov/products/datasets/watershed/*.

1. Introduction

1.1 Purpose

This document establishes Federal standards and procedures for creating a national, consistent, seamless, and hierarchical hydrologic unit dataset based on topographic and hydrologic features across the United States and territories. The WBD defines the perimeters of drainage areas (hydrologic units), formed by the terrain and other landscape characteristics, at a 1:24,000 scale in the United States, except for Alaska at 1:63,360 scale and 1:25,000 scale in the Caribbean, and it consists of digital geographic data that include six levels of detailed nested hydrologic unit boundaries.

This document serves as interagency guidance for the improvement of digital geographic data for 10-digit hydrologic units (Watersheds, formerly referred to as fifth level) and 12-digit hydrologic units (Subwatersheds, formerly referred to as sixth level), and for the optional delineation of 14- and 16-digit hydrologic units. The guidelines in this document are intended to provide a consistent framework for local, regional, and national needs in States, Tribal Lands, Pacific Islands, Puerto Rico, and the U.S. Virgin Islands to accurately delineate watersheds. This document sets forth terminology, definitions, and procedural information to ensure the uniform development of hydrologic unit boundaries and numerical codes by the agencies, tribes, and other organizations that develop, manage, archive, exchange, and analyze data by hydrologic features. The information presented is intended to enable users from different agencies and programs to contribute to coordinated watershed management, to efficiently share information and resources, and to ensure the digital geographic data are usable with other related Geographic Information System (GIS) data.

The use of these criteria and techniques for hydrologic unit selection and boundary delineation will permit use of the standardized hydrologic units by a diverse group serving multiagency programs. Some examples of these programs include watershed management, water-quality initiatives, watershed modeling, resource inventory and assessment, and Total Maximum Daily Load development. The usefulness of hydrologic units in a variety of sizes based on natural surface-water flow and topography cannot be overestimated for analytical and statistical purposes and applications. Instances of unusual landforms or artificial features that affect the hydrologic units are recorded in attributes and associated metadata, but in no way should they detract from the intent of these data to reflect surface-water flow.

The distribution of GIS layers and associated maps created by use of uniform guidelines and procedures improves the quality, consistency, and accessibility to hydrologic unit data nationwide, and thus becomes the framework for continued updates. Up-to-date information and availability of the hydrologic units are listed at *http://www.nrcs.usda.gov/wps/portal/nrcs/main/national/water/watersheds/dataset*. This document addresses concessions and changes necessary to meet database integration standards between the WBD and the National Hydrography Dataset (NHD). Updates to this edition also include additional content for instruction on the use of the optional higher resolution base data for improvements, stewardship, Geographic Names Information System (GNIS) integration, expanded domains, changes in the population of the modification field, and current information on international border harmonization. Of particular interest to the user community is the inclusion of offshore areas in coastal hydrologic units, which enhances vertical integration and eliminates duplicate features in the combined NHD and WBD geodatabase. Further enhancements include minimizing future coastline updates in the WBD for long-term stability and improving flexibility for raster applications and modeling needs.

This document refers to the nested hierarchy of the drainage areas by the hydrologic unit name (2-, 4-, 6-, 8-, 10-, and 12-digit hydrologic unit) rather than the historical names of the general classes of the hydrologic units (Region, Subregion, Basin, Subbasin, Watershed, and Subwatershed). This is compatible with WBD names integration with GNIS.

1.2   Background

A standardized system for organizing and collecting hydrologic data was developed in the mid-1970s by the U.S. Geological Survey (USGS) under the sponsorship of the Water Resources Council (Seaber and others, 1987). This system divided and subdivided the country into successively smaller hydrologic units based on surface features; established associated codes, names, and boundaries for the units; and classified them into four levels: Regions, Subregions, Accounting Units (later referred to as "Basins"), and Cataloging Units (later referred to as "Subbasins") (see section11.1). The hierarchical hydrologic unit code consists of two-digit numbers for each of the four nested hydrologic unit levels. These four levels aggregated form the 8-digit hydrologic unit code. The underlying hydrologic concept is a topographically defined set of drainage areas organized in a nested hierarchy by size and number of divisions per nested level.

The standardized 8-digit USGS Hydrologic Units are broadly used; however, the geographic area of the units was too large to adequately serve many water-resource investigations or resource-analysis and management needs. For example, the focus of many water-resource issues is pollutant loading and land-surface processes and the cumulative effects of pollution over space and time. Management of these issues required working with areas smaller than those defined by the 8-digit hydrologic units. Examples of programs requiring smaller hydrologic units include State river basin management plans; the U.S. Department of Agriculture (USDA), Natural Resources Conservation Service (NRCS) conservation and watershed programs; USDA Forest Service (USDA–FS) land-management-planning and watershed-management programs; various programs in the U.S. Environmental Protection Agency (USEPA), Office of Water; and programs in the USGS.

The NRCS is responsible for working with landowners to protect, improve, and sustain natural resources on private lands. In the early 1980s, the NRCS, which was known as the Soil Conservation Service (SCS) prior to 1994, completed mapping of 10-digit hydrologic units on small-scale State base maps for use in natural resource planning. In the mid-1990s, the NRCS, along with State agency conservation partners, began a national initiative to delineate and digitize 10- and 12-digit hydrologic units using methods and procedures that result in data that meet *national map accuracy standards*. In 1992, NRCS developed National Instruction 170–304 to promote a standardized criterion for hydrologic unit determination and delineation that would serve as the agency's policy for delineating and digitizing 10- and 12-digit hydrologic units (Soil Conservation Service, 1992). The NRCS updated the policy in 1995, incorporating changes from internal and external reviews. The NRCS has made considerable contributions to the development of nationally standardized 10- and 12-digit hydrologic unit boundaries (Natural Resources Conservation Service, 2002).

The USDA–FS and the Bureau of Land Management (BLM) are the primary land-management agencies of Federal lands in the United States. These agencies, and coordinating States, delineated and digitized 10- and 12-digit hydrologic units within an 8-digit hydrologic unit. Earlier delineations of hydrologic units of federally administered public lands served administrative purposes, but they were often developed without full coordination between Federal and State agencies.

The USGS and member agencies of the Federal Geographic Data Committee (FGDC), Subcommittee on Spatial Water Data (SSWD), coordinated and conducted a series of regional workshops to develop a nationally consistent hydrologic unit digital dataset. Member agencies of the SSWD assisted the NRCS and USGS in reviewing and certifying hydrologic units. Similarly, member agencies and others have assisted in researching techniques to employ digital elevation data for producing hydrologic unit delineations that function as draft lines to adjust to 1:24,000 contour lines where needed.

Although the BLM and USDA–FS are responsible for managing Federal lands, many State programs are responsible for managing State-owned land. As keepers of the land, States have partnered with Federal agency offices located within their State to develop a WBD that meets map accuracy standards and reflects local knowledge of surface-water resources.

A State agency is often the intermediary between local and Federal programs for land management. A representative example is North Dakota, where the Department of Health administers the nonpoint-source pollution control program. This watershed-based program provides EPA grant funds to local organizations for projects that reduce nonpoint-source pollution. In addition, the North Dakota Game and Fish Department uses the 12-digit hydrologic unit codes for their "Save Our Lakes" program, and the Private Lands Section of the Wildlife Division uses the 12-digit hydrologic unit codes for land-use inventories for their program.

To improve the sharing of national data, minimize the duplication of effort of various agencies, and discourage the creation of disparate datasets, the USGS and member agencies of the FGDC SSWD support a formal Memorandum of Understanding between the USGS National Geospatial Program and the NRCS National Cartography and Geospatial Center, now the National Geospatial Management Center (NGMC), signed in June of 2008. This agreement between the two agencies identifies the roles of the agencies and establishes the cooperative enhancement, maintenance, integration, and distribution of the WBD with the NHD and the inclusion of the WBD as a component of *The National Map*. The agreement benefits the cooperative development of the WBD and the NHD, as well as the business needs of the agencies, by making optimum use of an escalating amount of national watershed and hydrography data.

Since June of 2008, the USGS has started programmatic, data, editing tool, and stewardship integration of the WBD with the NHD. In 2012, USGS became the host of the WBD dataset of record, and NRCS is replicating this service. USGS and NRCS remain as copartners on the WBD.

2. Coordination

2.1 Federal Geographic Data Committee Subcommittee on Spatial Water Data

The SSWD was chartered and sponsored by the Advisory Committee on Water Information (ACWI) and the FGDC. The SSWD of the FGDC coordinates spatial water data and information activities among all levels of government and the private sector. Spatial water data include information about streams, hydrologic units, lakes, groundwater, coastal areas, precipitation, and other hydrologic information related to water resources.

Federal and State agencies involved in development and use of hydrologic units for water-resource management responsibilities are encouraged to participate as members of the SSWD. The SSWD assists the ACWI and FGDC by facilitating the exchange and transfer of water data; establishing and implementing standards for quality, content, and transfer of water data; and coordinating standards and collection of geographic data to minimize duplication of efforts.

2.2 National Steering Committee and WBD National Technical Coordinators

To facilitate development of nationally consistent data, two groups were established in the late 1990s. The first is the National Steering Committee, consisting of Federal representatives led by the NRCS and the USGS. The second is the WBD National Technical Coordinators (WBD–NTC). The groups engage in program management and long-term planning, integrate and coordinate with other national projects, and facilitate intrastate and interstate cooperation. The teams regularly clarify guidance, provide oversight and training to States, review interim State geographic data, suggest solutions for complex hydrographic landscapes, and review final data for standard compliance and incorporation into the WBD. The Steering Committee oversees final review and acceptance of digital geographic data into the national framework after recommendation by the WBD–NTC, grants certification, and thus approves the data for public access via the Web.

The USGS facilitates the WBD effort by maintaining the standards presented in this publication and by providing delineation support and review to all States. The WBD-NTC reviews data for standard compliance, and then the USGS integrates and merges the individual State certified datasets into a seamless national layer and processes the data for delivery on the *Geospatial Data Gateway.*

2.3 Intrastate and Interstate WBD Coordination

During creation of the WBD, further subdivision of the 8-digit hydrologic units into 10- and 12-digit hydrologic units provided an opportunity to develop consistent and commonly used nationwide digital geospatial data. States formed interagency hydrologic unit coordinating groups composed of Federal, State, local, and watershed agencies with an investment in developing hydrologic unit data, specifically including State agencies identified by State statute as having responsibility for the data. Each participating organization sought consensus appropriate to all interests and to obtain mutual technical approval. Each State coordinating group promoted the development, use, and maintenance of the WBD and identified a WBD In-State Steward to work with the WBD–NTC.

All States coordinated their delineation work within 8-, 10-, and 12-digit hydrologic units so that the digital geographic data matched across State borders. This coordination included the locations of outlet points, the sizes of hydrologic units, and the coding sequence within each level of the hydrologic unit hierarchy, as well as other attribute fields. The mapping of hydrologic boundaries across political boundaries was coordinated to ensure a nationally consistent dataset.

The interaction between these groups and data originators is shown in figure 1.

2.4 Stewardship of the WBD and the National Hydrography Dataset

The NHD is the surface-water component of *The National Map* and a companion dataset to the WBD. The NHD is a comprehensive set of digital spatial data that represents the surface water of the United States using common features such as lakes, ponds, streams, rivers, canals, stream gages, and dams. Polygons are used to represent area features such as lakes, ponds, and rivers, lines are used to represent linear features such as streams and smaller rivers, and points are used to represent point features such as stream gages and dams.

Although the WBD and NHD are administered under a separate model, they are coordinated so that a single source of universal, useful, and reliable hydrography and hydrologic units is produced and maintained. The improvement of both datasets focuses on spatial integration, application, and maintenance through stewardship.

Management of WBD–NHD data is distributed across the Nation, typically on a State-by-State basis. The WBD In-State Stewards coordinate and assume responsibility for identifying and implementing changes at the State level. Other organizations with specific local or topical interests may assume further stewardship under the auspices of the WBD In-State Stewards. In many cases, the State data stewards for the NHD and WBD are represented by different individuals and organizations.

The NHD approach to data stewardship provides documented roles and responsibilities between the USGS and NHD In-State Stewards. In most cases, existing NHD agreements will be modified to include WBD responsibilities, and new agreements will integrate stewardship for both datasets.

Because the WBD has national consistency standards, the WBD–NTC will continue to work closely with the WBD In-State Stewards. This will ensure that future revisions to the WBD follow the guidelines set forth in this document. Data revisions will become transactions provided for inclusion in the NHD geodatabase administered by the USGS on behalf of the entire user community. The interaction between the WBD and NHD data stewards after local enhancements are made is shown in figure 1.

Editing tools that allow data stewards to upgrade the individual datasets while preserving the integrity of the model and the geometric relationships in the hydrography datasets are the key to the implementation of data stewardship. Suites of editing tools are available for the NHD and the WBD. The USGS National Geospatial Technical Operations Center (NGTOC) is responsible for maintaining and improving these tools. The NHDGeoEdit tools work with a spatial geodatabase. To maintain the hydrologic units of the WBD in the future, the In-State Stewards will be provided with a similar, but unique, WBD tool set that facilitates editing WBD data and managing metadata.

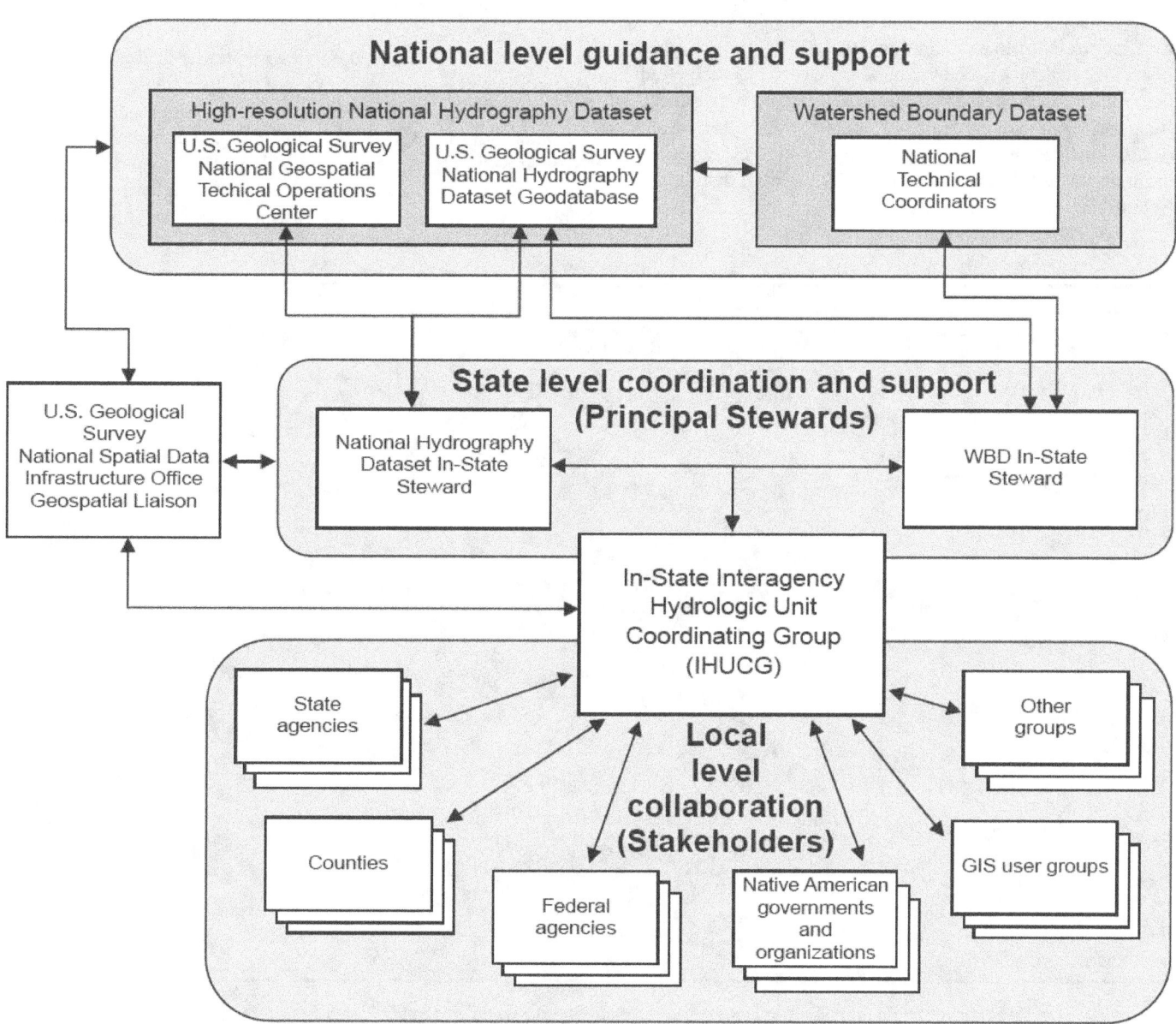

Figure 1.   Watershed Boundary Dataset coordination participants and relationships. (GIS, geographic information system)

3.   Criteria and Considerations for Delineating Hydrologic Units

3.1   General Criteria

This section describes criteria for determining and delineating 10- and 12-digit hydrologic units and optional 14- and 16-digit conventionally delineated hydrologic units. The 2- through 8-digit hydrologic units will be consistent with the new hydrologic units to create one dataset of 2- through 12-digit hydrologic unit codes and boundaries at a 1:24,000 scale in the United States, except for Alaska at 1:63,360 scale and 1:25,000 scale in the Caribbean. Selecting and delineating hydrologic units requires sound hydrologic judgment and must be based solely upon hydrologic principles to ensure a homogeneous, national, and seamless digital data layer. The diversity of hydrologic conditions nationwide, the complexity of surface hydrology, and the number of factors involved in the delineation process preclude an all-encompassing guideline. Variations will generally be limited to unusual hydrologic or landform features and dissimilar hydrologic or morphologic drainage-area characteristics. The intent of defining hydrologic units is to establish a baseline that covers all areas.

At a minimum, the hydrologic units must be delineated and georeferenced to USGS 1:24,000-, 1:25,000-, or 1:63,360-scale topographic maps, which meet the National Standard for Spatial Data Accuracy (NSSDA; *Federal Geographic Data Committee, 1998b*) (fig. 2). Digital Orthophoto Quarter Quadrangles (DOQQs) or higher resolution imagery contours of elevation, stream locations, and other relevant data are useful for keeping the hydrologic unit boundaries accurate and current.

The delineation must be as simple as is practical and avoid creating hydrologic units that favor a particular agency, program, administrative area, or special project. Drainage boundaries generated for special purposes that do not follow these guidelines will not be accepted into the WBD.

3.2   Hydrologic Boundaries

Figure 3 illustrates the United States hydrologic unit boundaries, the six nested levels of the hydrologic unit hierarchy, and the aggregated sequence of hydrologic unit codes (from a 2-digit to a 12-digit number). The large geographic area of 8-digit hydrologic units is the basis for subdividing 10-digit hydrologic units, and the 10-digit hydrologic unit boundaries are the basis for further subdividing 12-digit

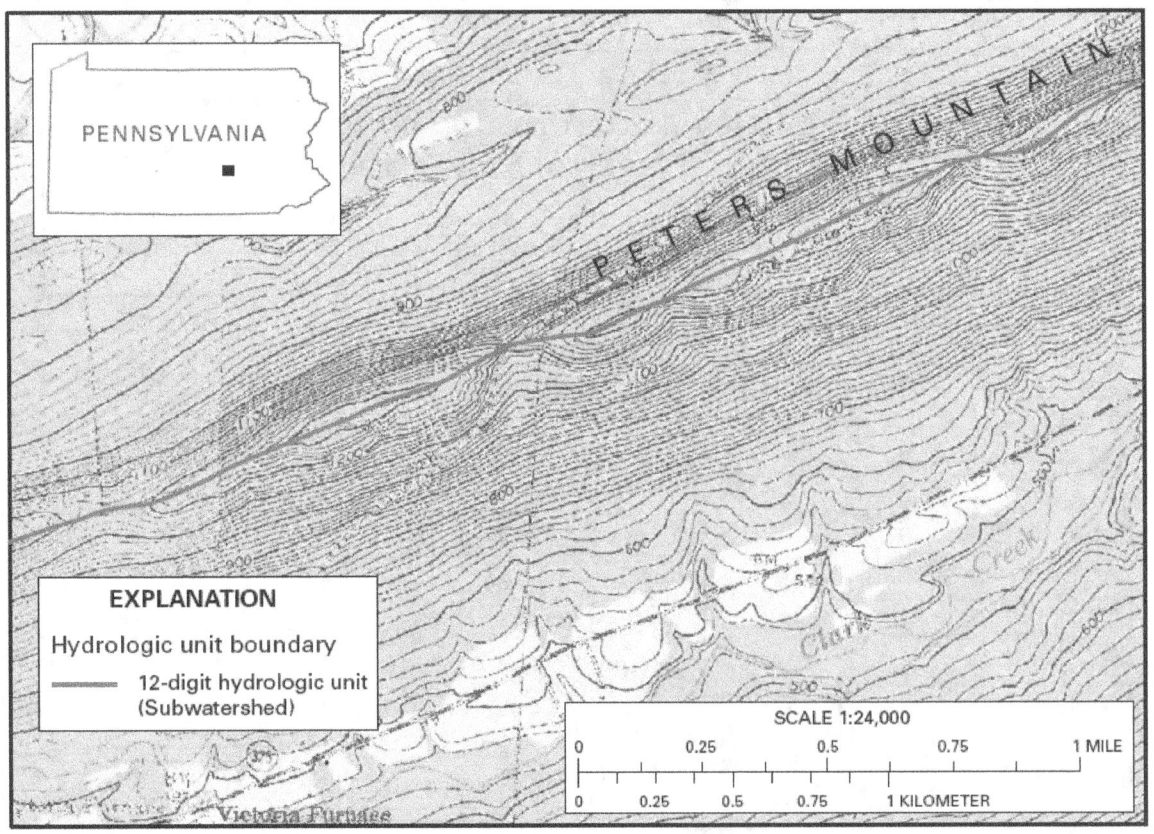

Figure 2.   This 12-digit hydrologic unit boundary meets the National Standards for the Spatial Data Accuracy and the Watershed Boundary Dataset standards because it is georeferenced to the minimum 1:24,000-scale topographic base map. The Appalachian Trail south of Enders, in Dauphin County, Pennsylvania, is shown.

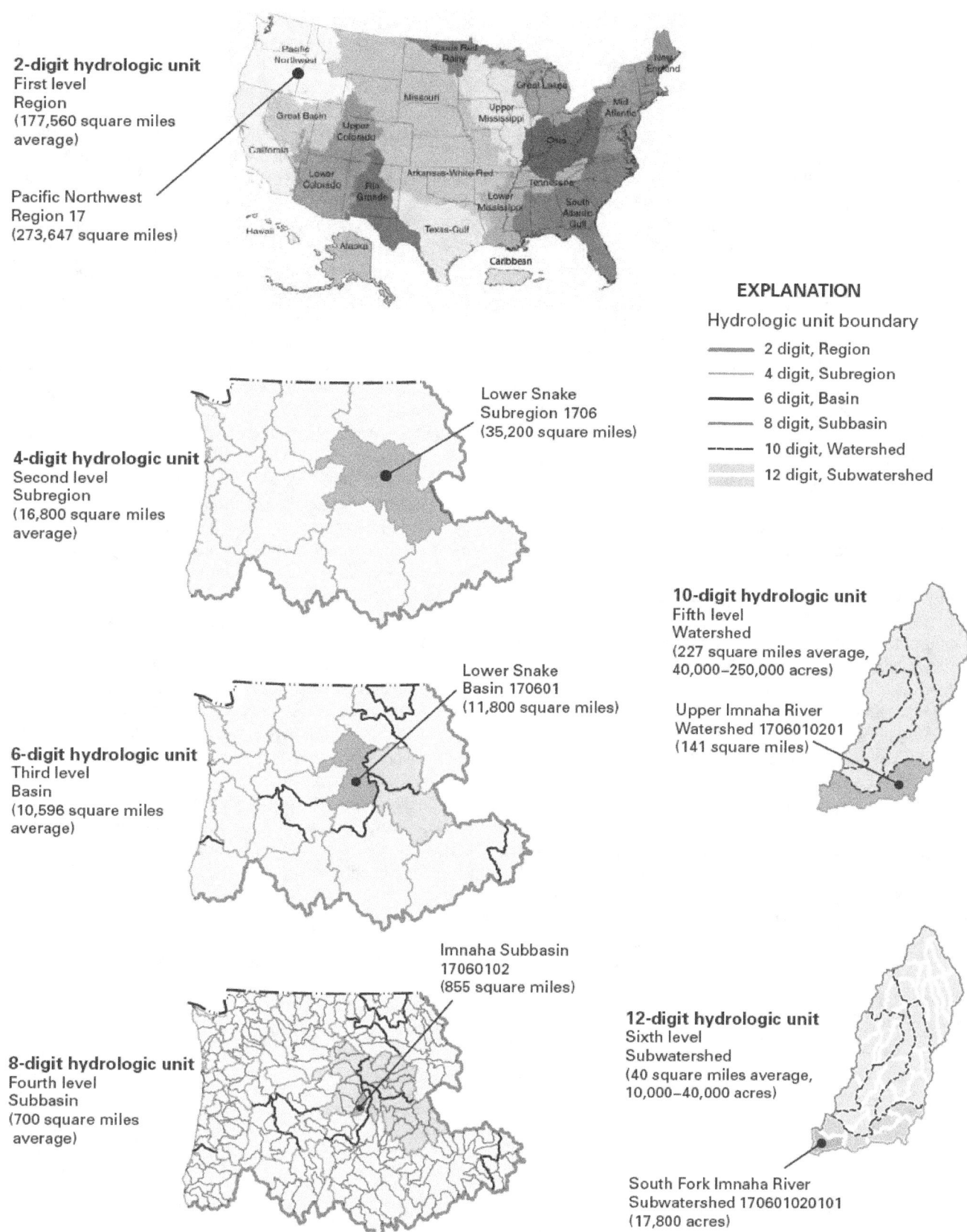

2-digit hydrologic unit
First level
Region
(177,560 square miles average)

Pacific Northwest Region 17
(273,647 square miles)

4-digit hydrologic unit
Second level
Subregion
(16,800 square miles average)

Lower Snake
Subregion 1706
(35,200 square miles)

EXPLANATION

Hydrologic unit boundary

2 digit, Region
4 digit, Subregion
6 digit, Basin
8 digit, Subbasin
10 digit, Watershed
12 digit, Subwatershed

10-digit hydrologic unit
Fifth level
Watershed
(227 square miles average, 40,000–250,000 acres)

Upper Imnaha River
Watershed 1706010201
(141 square miles)

Lower Snake
Basin 170601
(11,800 square miles)

6-digit hydrologic unit
Third level
Basin
(10,596 square miles average)

Imnaha Subbasin
17060102
(855 square miles)

12-digit hydrologic unit
Sixth level
Subwatershed
(40 square miles average, 10,000–40,000 acres)

8-digit hydrologic unit
Fourth level
Subbasin
(700 square miles average)

South Fork Imnaha River
Subwatershed 170601020101
(17,800 acres)

Figure 3. Hierarchy and areas for the six nested levels of hydrologic units are shown in the above example. As they are successively subdivided, the numbering scheme of the units increases by two digits per level.

hydrologic units. By delineating 10- and 12-digit hydrologic units at a 1:24,000 scale in the United States, except for Alaska at 1:63,360 scale, and 1:25,000 scale in the Caribbean, line precision and accuracy will be propagated throughout the dataset.

If more accurate base reference information becomes available and editing suggests that more than minor changes are needed to the 8-digit hydrologic unit boundary, then the proposed edits must be coordinated with and preapproved by the WBD–NTC. If changes or revisions to 8-digit hydrologic unit boundaries are needed in addition to resolution improvements required for 1:24,000-scale data (section 4.4), then inform the WBD In-State Steward or interagency hydrologic unit group point of contact of the major change to prompt consultation and direction from the WBD–NTC.

The 10- and 12-digit hydrologic units, like other hydrologic units, are defined along natural hydrologic breaks based on land surface and surface-water flow. Boundaries delimit the land area of the 10- and 12-digit hydrologic units. A hydrologic unit has a single flow outlet except in frontal, lake, braided-stream, or playa (closed basin) hydrologic units (sections 3.5, 3.6). A hydrologic unit with an outlet at a delta or braided stream should be treated in a similar manner to that of a frontal hydrologic unit. Give priority to delineating 10- and 12-digit hydrologic units that will be "standard" units having only one outlet (section 3.5.1). Because 10- and 12-digit hydrologic units are subdivisions of a higher level of hydrologic unit, they must share common boundaries with the existing hydrologic units defined in higher levels of the hydrologic unit hierarchy. A dam, diversion, or stream confluence may be used to divide a hydrologic unit into upper and lower parts.

Previous versions of 10- and 12-digit hydrologic unit boundaries may have used administrative boundaries as hydrologic unit boundaries. Now and in the future, hydrologic unit boundaries must be defined solely by examination of topography and hydrologic features. Do not use administrative or political boundaries, such as county, State, or national forest boundaries, as criteria for defining a hydrologic unit boundary unless the administrative boundary is coincident with a topographic feature that appropriately defines the hydrologic

unit. Existing hydrologic unit data that include boundaries delineated solely by use of administrative or political boundaries will not be certified as meeting these guidelines until the hydrologic units are revised based on topography, surface-water flow, and hydrologic features.

Hydrologic unit delineations along the international boundary of the United States are to be coordinated through the WBD–NTC because established processes, international efforts, and identified Federal partners are in place with Canada and Mexico. For units including offshore waters, use the National Oceanic and Atmospheric Administration (NOAA) Three Nautical Mile Line (section 3.6) as the offshore limit.

Boundary delineations based on hydrology include land areas on both sides of a stream flowing toward a single downstream outlet, except in the case of open-water hydrologic units (section 3.6). Boundaries should not follow or run parallel to streams as delineated hypothetically in figure 4A, except where physical features such as levees, berms, incised channels, and similar structures prevent water from flowing directly to the outlet. Do not delineate boundaries down the middle of a stream. Figure 4B illustrates correct delineation. Boundaries will cross the stream perpendicular to flow at the hydrologic unit outlet (fig. 5). Where the main stem hydrologic unit outlet is defined at a confluence with a major tributary, the boundary should be placed on the downstream side of the confluence (fig. 5). The hydrologic unit boundary may use smaller tributaries as the delineation point to divide the hydrologic unit into suitably sized 10- or 12-digit hydrologic units. Delineating the boundary at a confluence accommodates the nesting of smaller units within the hydrologic unit for future site-specific planning, assessment, monitoring, or inventory activities. Information from stream gages, locations of major highway crossings, and NHD reach endpoints may aid in the identification of hydrologic unit divisions.

In addition to the primary criteria, there are general criteria for the number of hydrologic units subdivided from a higher level unit, the size of hydrologic units, and the treatment of noncontributing and remnant areas.

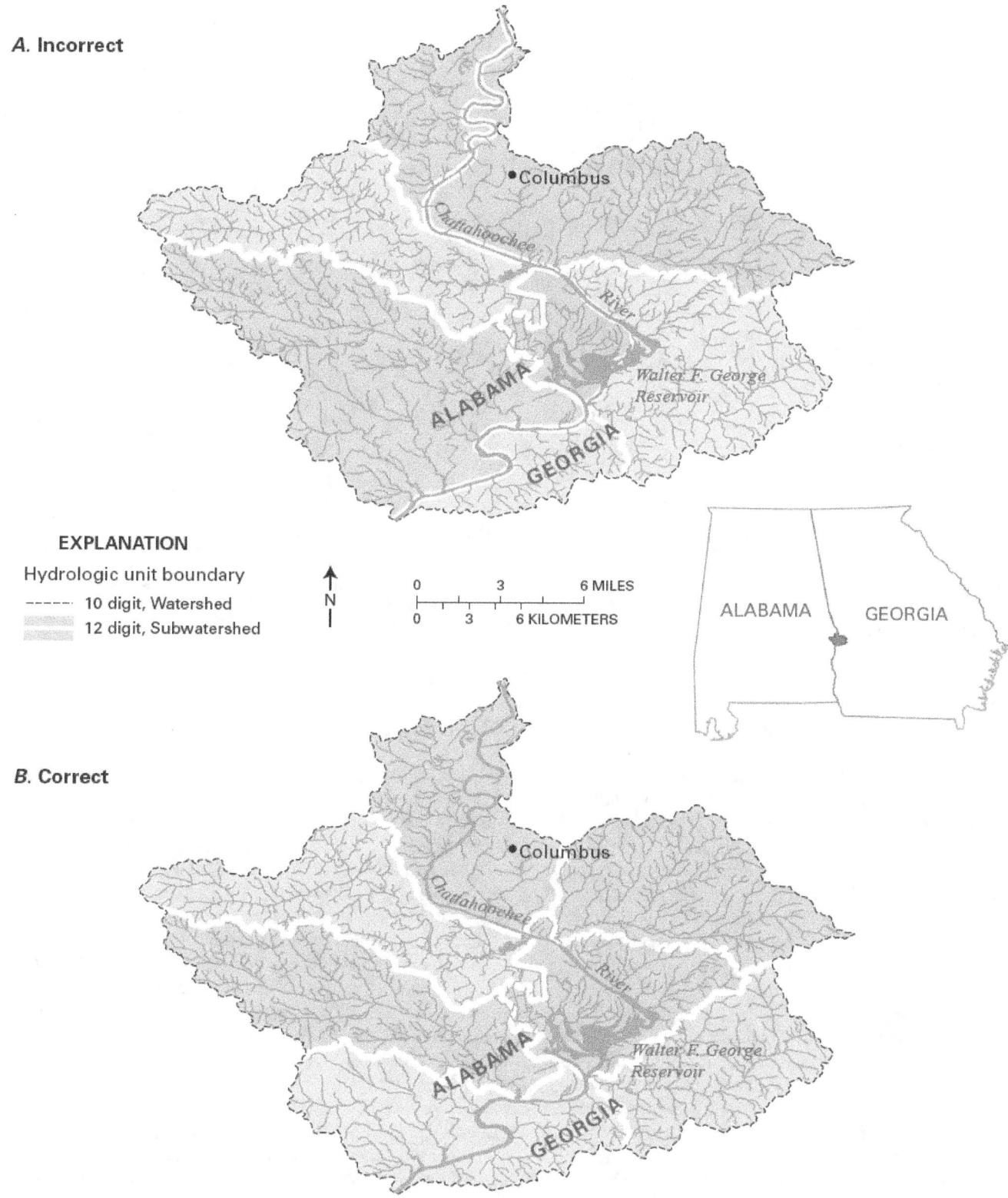

Figure 4. Delineation of hydrologic features must be based on land surface, surface-water flow, and hydrographic features. *A.* Incorrect—8-digit hydrologic unit boundaries follow the river and administrative boundaries. *B.* Correct—8-digit hydrologic unit boundaries cross the Chattahoochee River and end at the bank of the receiving stream for tributaries. The area includes the Walter F. George Reservoir south of Columbus, Georgia.

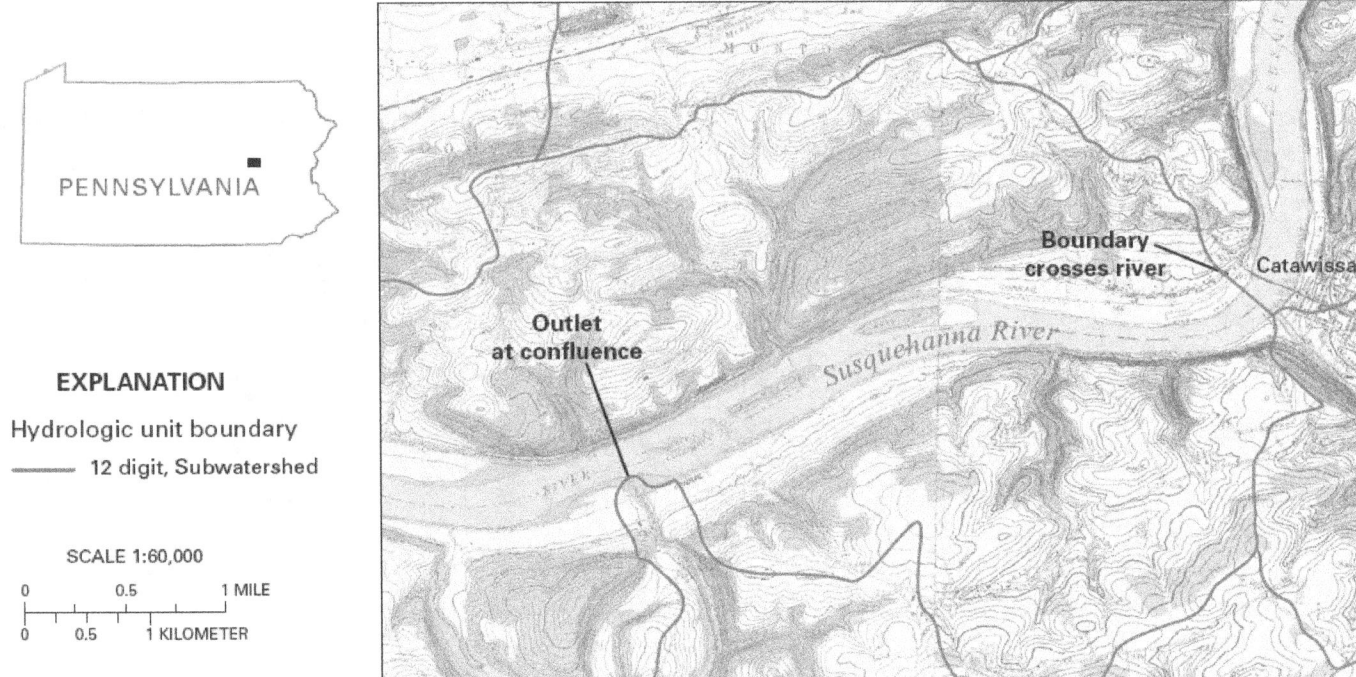

PENNSYLVANIA

EXPLANATION

Hydrologic unit boundary

――――― 12 digit, Subwatershed

SCALE 1:60,000

Figure 5.   The two tributary outlets shown above are delineated at the bank of the receiving stream. The boundary correctly crosses the Susquehanna River perpendicular to the river just downstream from the tributary confluence. The area is south of Bloomsburg, in Columbia County, Pennsylvania.

3.3   Number of Subdivided Hydrologic Units

As a general rule, subdivide each hydrologic unit into 5–15 units. For example, 5–15 10-digit hydrologic units will be nested in each 8-digit hydrologic unit. This system accommodates geomorphic or other relevant basin characteristics and creates a fairly uniform size distribution of same-level hydrologic units within a broad physiographic area. This system also results in a smooth transition between sizes of same-level hydrologic units as topography changes between physiographic areas and maintains consistent delineations across State borders.

There are exceptions to using the 5–15 rule, just as there are existing 4-digit hydrologic units that contain fewer than five 6-digit hydrologic units and 6-digit hydrologic units that contain fewer than five 8-digit hydrologic units. The number of 10-digit hydrologic units nested within 8-digit hydrologic units or the number of 12-digit hydrologic units nested within some 10-digit hydrologic units may occasionally be reduced or expanded to accommodate special areas. In some places, it is not possible to delineate 10- and 12-digit hydrologic units owing to the lack of hydrologic features or insufficient topography. Where offshore boundaries are required, 8-digit hydrologic units must be extended to the NOAA Three Nautical Mile Line.

3.4   Sizes of 10- and 12-Digit Hydrologic Units

The hydrologic units of any given level should be about the same size within a physiographic area for comparability and coordinating use and applications of hydrologic units. Hydrologic units should not be substantially different in size from the rest of the hydrologic units for a given level.

Nationally, the typical size for a 10-digit hydrologic unit is 40,000–250,000 acres. Use this acreage range as a guide in subdividing 8-digit hydrologic units. Each 10-digit hydrologic unit must be completely contained within one 8-digit hydrologic unit. The typical size for a 12-digit hydrologic unit is 10,000–40,000 acres; however, in some areas with unique geomorphology the 12-digit hydrologic units may be greater than 40,000 acres or less than 10,000 acres, but never less than 3,000 acres. A variance outside the size criteria of up to 10 percent of the polygons within a State is allowed for both the 10-digit and the 12-digit hydrologic unit.

In coastal areas where radial or centripetal drainage predominates, such as Hawaii, individual streams with outlets to the ocean, or remnants, may be less than 3,000 acres each. Their acreage can be combined into a single hydrologic unit of greater than 10,000 acres. In nonterrestrial coastal hydrologic units, subdivision to meet size criteria is not required, however the benefit of uniformity of sizes at any level should be considered.

3.5   Geomorphic Considerations for Hydrologic Units

This section explains the most common geomorphic circumstances and details those that require additional considerations when hydrologic units are developed.

3.5.1   Standard Hydrologic Units

The standard hydrologic unit is defined by the surface-water drainage area (fig. 6). All of the surface drainage within the standard hydrologic unit boundary converges at a single outlet point. Larger standard hydrologic units may be subdivided into multiple hydrologic units in order to fit within the size criteria of a given level.

Delineate by starting from the designated outlet (a point on a single stream channel that drains the area), and proceed to the highest elevation of land dividing the direction of water flow. This boundary connects back to the designated outlet, where it will cross perpendicular to the stream channel. Correctly selecting the outlet point is critical to delineating all hydrologic units accurately. The standard hydrologic units are subdivisions of higher level hydrologic units based on major tributaries. When choosing between tributaries for hydrologic unit delineation, larger streams are typically chosen over smaller streams. The downstream end of the hydrologic unit will be at the confluence with the main stem of the higher level hydrologic unit or the main stem of a hydrologic unit of the same level, when possible (fig 5.).

A hydrologic unit may be divided at a lake outlet if the upstream drainage-area size is appropriate for the hydrologic unit level being delineated. The boundaries should be as simple as possible while capturing the topographically defined area that contributes to the outlet.

Identifying standard hydrologic units and delineating them with the recommended number of subdivisions and area sizes will cover much of the area of the higher level hydrologic unit to be subdivided. Where nonstandard areas such as remnant (section 3.5.2), noncontributing (section 3.5.3), and diverted (section 3.5.4) areas exist, they will need to be delineated by use of criteria described in the following sections. Nonstandard areas are typically added to adjacent hydrologic units, but they occasionally may have to exist as small, atypical hydrologic units.

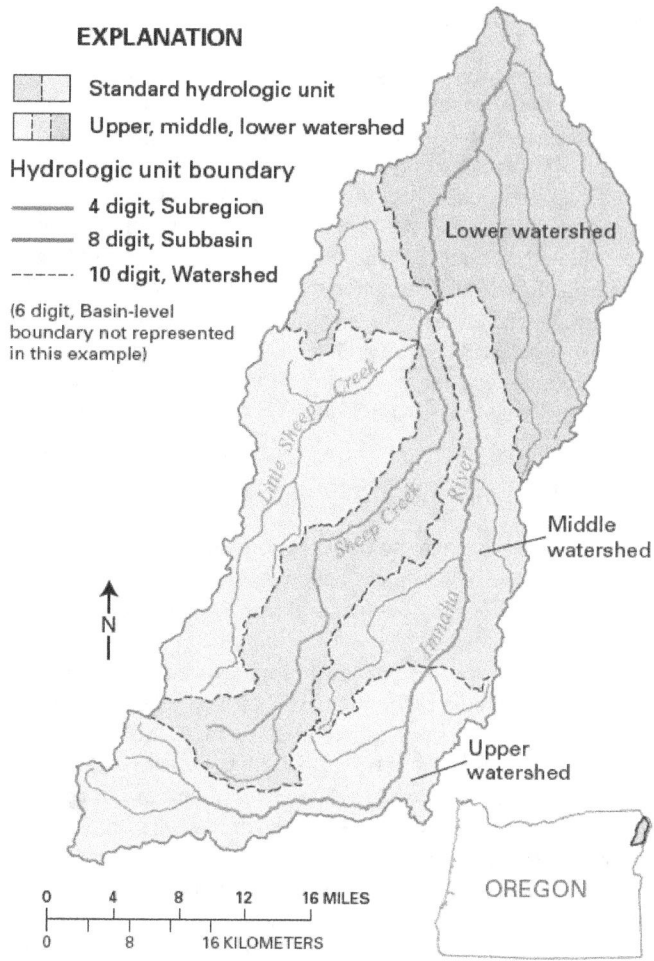

EXPLANATION

Standard hydrologic unit

Upper, middle, lower watershed

Hydrologic unit boundary

——— 4 digit, Subregion

——— 8 digit, Subbasin

- - - - 10 digit, Watershed

(6 digit, Basin-level boundary not represented in this example)

Figure 6.   In this example of an 8-digit hydrologic unit defined by the hydrology (Imnaha Subbasin), the main-stem stream has been split into "Upper, "Middle," and "Lower" watersheds and its tributary watersheds. The Imnaha River is in Wallowa County, Oregon.

3.5.2   Remnant Areas

Delineating hydrologic units may result in remnant areas around the main stem of larger streams, even when sound hydrologic judgment and standard practices described above are used. Remnant areas typically occur as wedge-shaped areas along interfluvial regions between adjacent standard hydrologic units or as overbank areas along a stream between junctions with tributaries. These remnant areas, along with the mainstem, should be aggregated together into hydrologic units that meet the size criteria for any given level. These remnant areas are also referred to as "related contributing drainage areas" or "composite" areas. See section 3.6 for discussion of nonstandard areas along shorelines.

3.5.3   Noncontributing Areas

Drainage areas that do not contribute flow toward the outlet of a hydrologic unit are called noncontributing areas (fig. 7). Such areas may be due to glaciated plains (potholes), closed basins, playas, cirques, depression lakes, dry lakebeds, or similar landforms. A noncontributing area may be desig- nated as a hydrologic unit at any level of the hierarchy if it is within the size range for a given level. Semiconfined basins that contribute surface water to other areas in wet years but act as sinks in dry years may be defined as standard or noncon- tributing hydrologic units. These types of special situations require review, coordination, and agreement at the State level. Assistance or consultation with climatologists or NOAA on prevailing precipitation regimes that may have a long-term influence on noncontributing areas should be explored.

If noncontributing areas are small and dispersed relative to the hierarchical level being delineated, then they should be considered as part of the encompassing delineated hydrologic unit. Because the precise definition of a noncontributing area will likely vary from State to State, document the criteria used to determine noncontributing areas in the metadata file, especially if a substantial number of noncontributing areas are defined. Include the total acreage of noncontributing areas within a hydrologic unit as an attribute of the data. Delineate noncontributing areas consistently. If a noncontributing area is on the boundary between two or more hydrologic units, then determine the low point along the noncontributing area bound- ary, and associate the noncontributing area with the hydrologic unit adjacent to the low point on the boundary.

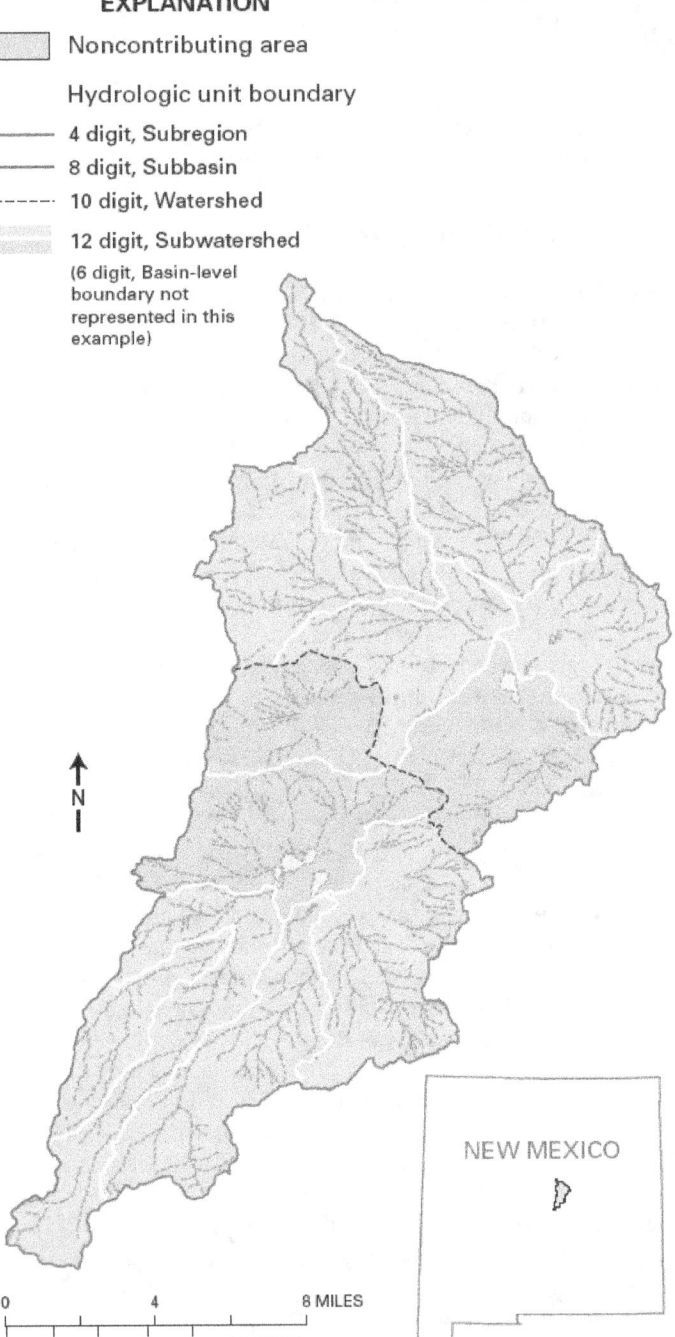

EXPLANATION

▨ Noncontributing area

Hydrologic unit boundary

——— 4 digit, Subregion

——— 8 digit, Subbasin

----- 10 digit, Watershed

▦ 12 digit, Subwatershed

(6 digit, Basin-level boundary not represented in this example)

↑ N

NEW MEXICO

0   4   8 MILES

0   4   8 KILOMETERS

Figure 7.   Noncontributing areas, caused by unique and unusual landforms, do not flow to the outlet of the hydrologic unit. Isolated noncontributing areas such as these in Eastern Estancia Subbasin, Torrance County, New Mexico, should be delineated.

3.5.4 Diverted Waters

Ditches and canals should be used to determine surface-water drainage areas only when the artificial channel has permanently altered the natural flow. Many artificial drainage features in the United States were originally either perennial or intermittent channels that local government and private entities converted into permanent drainage features. Much of the surface drainage in these areas would disappear from local and State drainage maps if permanent, constructed diversions were not considered when delineating hydrologic units. If the present-day canal or ditch was once a legacy stream channel or has perennial flow, then it may be considered for delineating hydrologic units. Avoid delineating small, local ditch systems constructed for seasonal diversion of water or for irrigation of agricultural fields.

When all or part of the flow from one hydrologic unit is continuously discharged into another by constructed transbasin diversions, document the diverted flow as attributes to both the water-losing and water-gaining hydrologic units in the polygon 12-digit HUMod field (section 6.2. 2.11.2).

Information on the date of the diversion, flow rates, and water rights for both the receiving and losing hydrologic units may be included in the metadata.

Avoid adjusting the location of hydrologic units because of interconnected flow from one hydrologic unit to another during high flow stages in streams.

3.5.5 Karst Areas

The delineation of hydrologic unit boundaries in karst areas is difficult because the potential exists to attribute the runoff from one hydrologic unit, or part of a hydrologic unit, to the wrong outlet. This difficulty occurs because the surface drainage pattern in karst areas is typically disrupted by sinkholes and sinking streams, making it difficult to choose a valid hydrologic unit boundary based on topography alone. Karst conduits frequently cross beneath topographically defined hydrologic unit boundaries, and drainage may be routed into or out of a topographically defined hydrologic unit. Because of the presence of karst conduits, groundwater and surface-water interaction is relatively direct and rapid compared to that in nonkarst areas. The conduit networks transport most of the stormwater runoff, and the flow is both fast and turbulent. Each conduit network drains a finite area (karst basin) and discharges to a perennial spring.

Sinkholes, sinking streams, springs, or cave entrances indicate karst hydrology. If sinkhole symbols or other map symbols associated with karst areas are not expressed on topographic maps that show areas with soluble rocks such as limestone, dolomite, gypsum, or salt near the land surface, then assume karst hydrology unless evidence indicates otherwise. The WBD is a surface-water dataset; it is not the intent of the WBD to determine or delineate in depth the pattern of underground seepage. Note all hydrologic units containing karst with the modifier "KA" in the relevant 12-digit hydrologic unit polygon HUMod field (section 6.2. 2.11.2).

Additional details describing interpretation or identification of karst within the hydrologic unit should be documented in metadata (section 9).

3.6 Coastal Considerations for Hydrologic Units

The development of hydrologic units in coastal areas requires additional consideration, because coastal units may drain either to a single outlet (standard) or to multiple outlets (frontal drainage). Beginning in 2009, the shoreline feature was not included as a component of the WBD; therefore, an offshore boundary is now required for representing the outer extent of the hydrologic unit for an open ocean coastal area and an onshore topographically defined boundary is required for inland coastal areas. Nevertheless, the shoreline representation is retained within the NHD as part of the Hydrography geodatabase for areas such as inland lakes, large tidal rivers such as parts of the Mississippi, Columbia, and Potomac Rivers, and for ocean coastal areas. If the NHD shoreline is in conflict with the established State shoreline representation, then contact your In-State Steward to determine which version should be incorporated into the NHD.

The offshore boundary can be generally categorized as open-ocean boundary or nearshore boundary. Open-ocean hydrologic units are closed by use of the nationally consistent NOAA Three Nautical Mile Line, as published on the NOAA nautical chart. Charts can be accessed at *http://www. nauticalcharts.noaa.gov/csdl/mbound.htm*. For digital representation of this demarcation, contact the WBD–NTC. Nearshore delineation is optional.

To eliminate the shoreline in bays, sounds, and estuaries, use criteria established by the State; these criteria may include the nearshore boundary delineation or the combination of a frontal and a water unit closed by a head of land boundary delineated across the mouth. The boundary should be far enough offshore that it does not interfere with other shoreline representations, for example, mean low water.

Extend the boundaries of frontal coastal hydrologic units to intersect offshore boundaries. When the NOAA Three Nautical Mile Line is the outer extent, 8- and 10-digit hydrologic units must be delineated across open ocean, preferably along submerged ridges. Nonstandard units that flow directly into coastal waters may require extension to an offshore boundary (fig. 8).

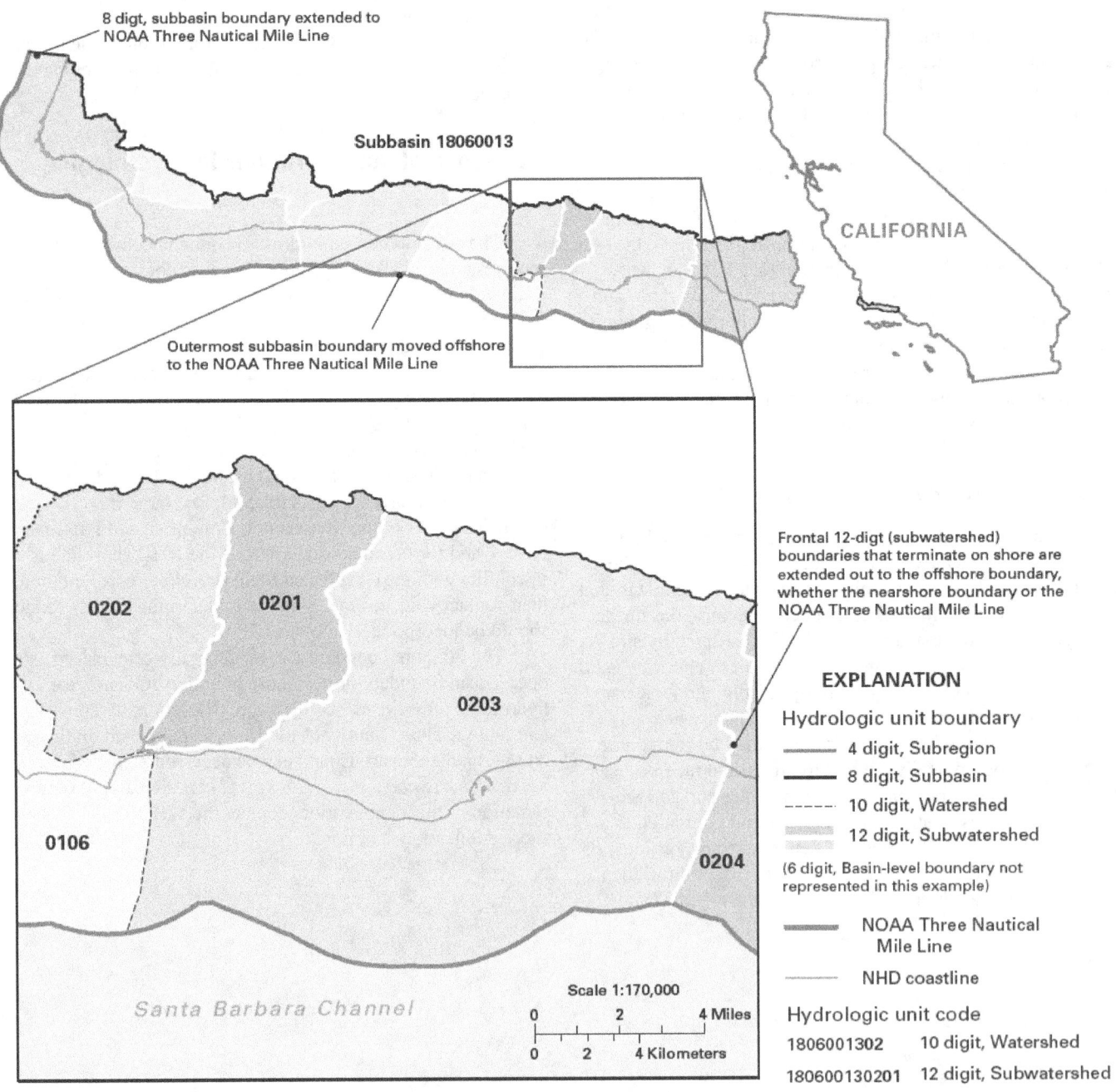

Figure 8.   To delineate the hydrologic unit boundary for frontal hydrologic units in open ocean areas, extend the boundaries of the frontal unit to intersect offshore boundaries. (NHD, National Hydrography Dataset; NOAA, National Oceanic and Atmospheric Administration)

3.6.1 Bays, Sounds, and Estuaries

The delineation of hydrologic unit boundaries in coastal areas is often complicated by the presence of large estuaries, bays, or sounds. Use the NHD to identify and define bays, sounds, and estuaries. Because coastline representations are not stored within hydrologic unit boundaries, either nearshore boundary delineation or a combination frontal and water (section 3.6.2) unit closed by a head of land boundary (delineated across the mouth) is required, as agreed upon by State partners (fig. 9). Water levels in coastal areas can fluctuate substantially; therefore, consideration should be given to using either a nearshore buffer distance or the bathymetry of submerged morphologic features, such as stable shoals, ridges, shore faces, and flow channels (legacy channels), if the bathymetry is reliable and current. Transitory features, such as berms that are created by sediment flow from a tributary and later destroyed by other action, should not be used. Any boundary

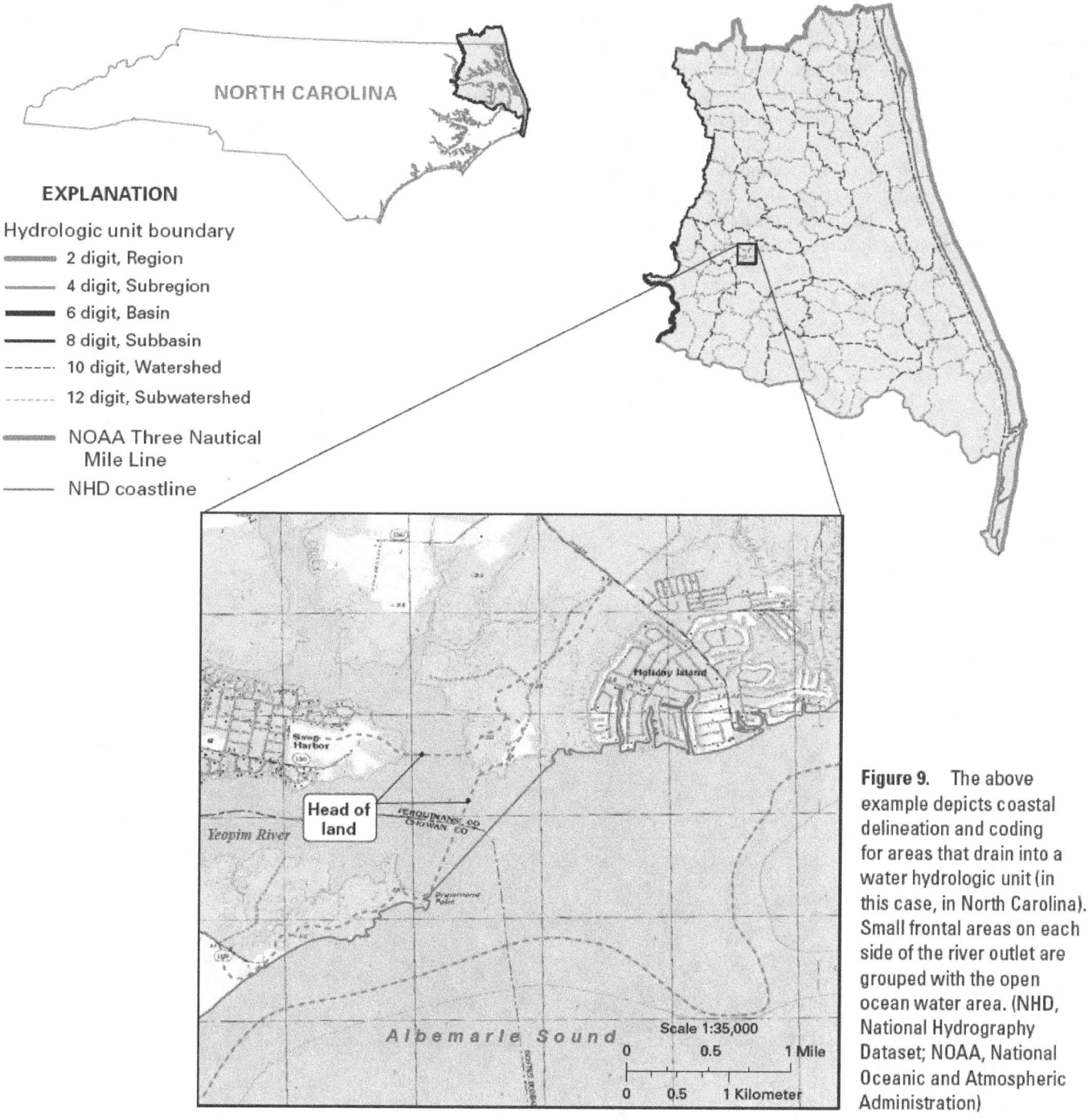

EXPLANATION

Hydrologic unit boundary

- 2 digit, Region
- 4 digit, Subregion
- 6 digit, Basin
- 8 digit, Subbasin
- 10 digit, Watershed
- 12 digit, Subwatershed

- NOAA Three Nautical Mile Line
- NHD coastline

Figure 9. The above example depicts coastal delineation and coding for areas that drain into a water hydrologic unit (in this case, in North Carolina). Small frontal areas on each side of the river outlet are grouped with the open ocean water area. (NHD, National Hydrography Dataset; NOAA, National Oceanic and Atmospheric Administration)

chosen for delineation should be far enough offshore so that it will not interfere with other shoreline representations, for example, mean low water.

To delineate nearshore hydrologic unit boundaries, a combination of depth (bathymetry), head of land, and variable distances (buffering) from the NHD coastline representation may be applied or frontal land and water units may be combined (fig. 10). When the nearshore boundary is a buffer distance from a shoreline, the buffer distance may differ from the shoreline representation in the NHD, which typically is the Mean High Water tidal datum supplied by NOAA. If the NHD shoreline representation is in conflict with the authoritative State shoreline, then contact your NHD In-State Steward. When the nearshore delineation is based on a depth, it may vary, for example, from 6 to 10 feet. When long sinuous estuaries or rivers are delineated, three options may be used: (1) truncate the river or estuary with a head of land boundary delineated across the mouth to separate the inland bay from the water unit (fig. 9); (2) reduce the buffer distance at the point where the buffers overlap, for example, at a narrow channel at the mouth of a bay; or (3) create an artificial closure line beyond the mouth at the outer buffer extent.

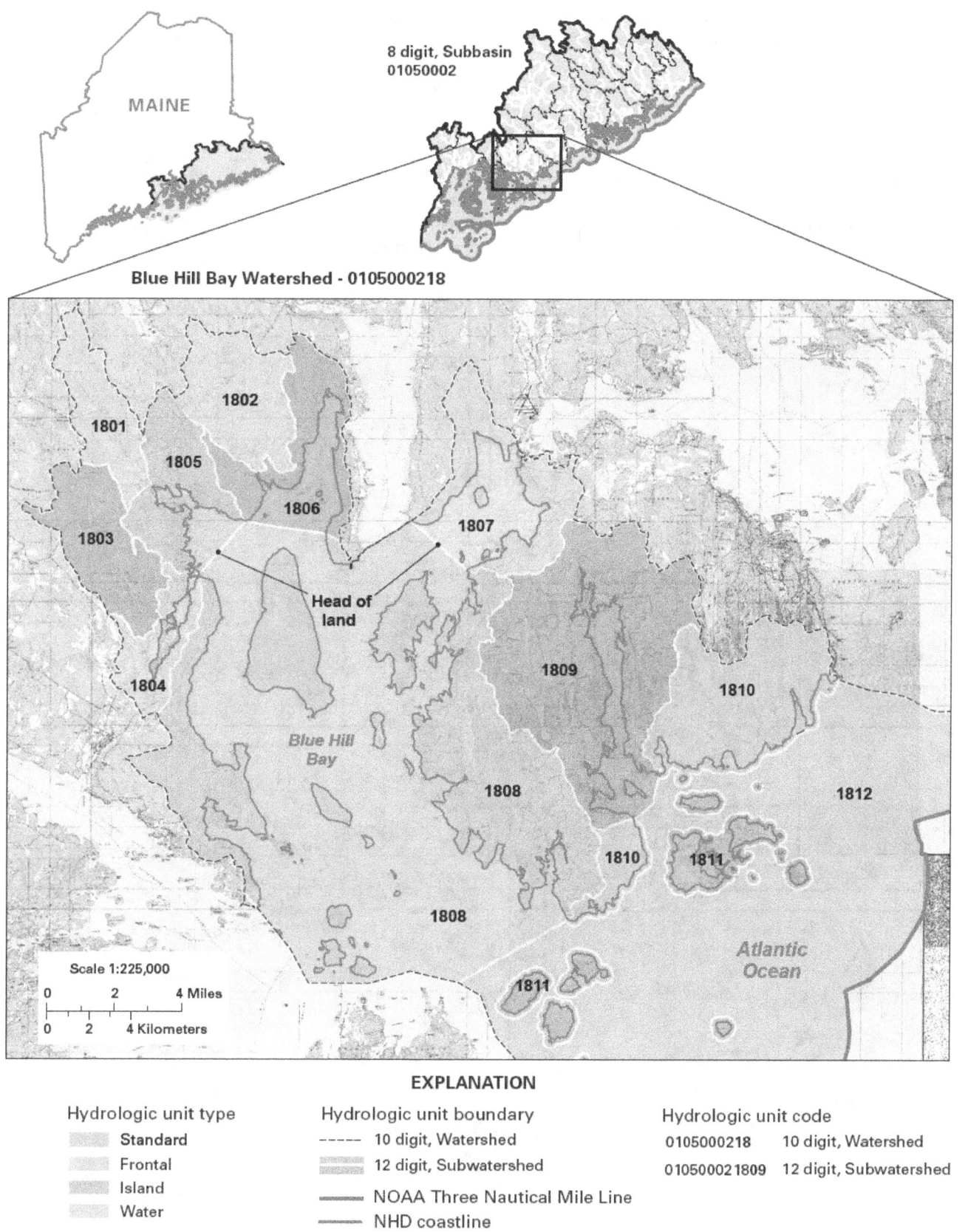

Blue Hill Bay Watershed - 0105000218

EXPLANATION

Hydrologic unit type
- Standard
- Frontal
- Island
- Water

Hydrologic unit boundary
- - - - -   10 digit, Watershed
          12 digit, Subwatershed
————   NOAA Three Nautical Mile Line
————   NHD coastline

Hydrologic unit code

0105000218   10 digit, Watershed

010500021809   12 digit, Subwatershed

Figure 10.   A combination of depth (bathymetry), head of land, and variable distances (buffering) from the NHD coastline representation may be applied or frontal land and water units may be combined when nearshore hydrologic unit boundaries are delineated. (NHD, National Hydrography Dataset; NOAA, National Oceanic and Atmospheric Administration)

When delineating from bathymetric information, use the best available depth data for closure as confirmed by State partners. Extend the boundaries of coastal hydrologic units to intersect offshore boundaries. Delineations of 8-digit hydrologic units must be extended to nearshore and open-ocean boundaries, preferably along submerged ridges. Where bathymetry is not available, one must project an arbitrary perpendicular line out to the offshore boundary.

Nautical charts from NOAA or local marine bathymetric data are available for many coastal areas. Help in obtaining bathymetric data is available through the WBD–NTC. Use the largest scale and most recent charts and maps that provide individual depth-sounding and depth-contour data. Digital elevation model data may be useful for identifying water hydrologic units.

Nearshore delineations must transition seamlessly across State boundaries and physical features having edges that match. This may require negotiation between States to determine the transition from one nearshore representation to another. Nearshore delineations in bays, sounds, and estuaries that are adjacent to nearshore delineations in open-ocean areas may require a gradual transition between the two offshore depths/distances (fig. 11).

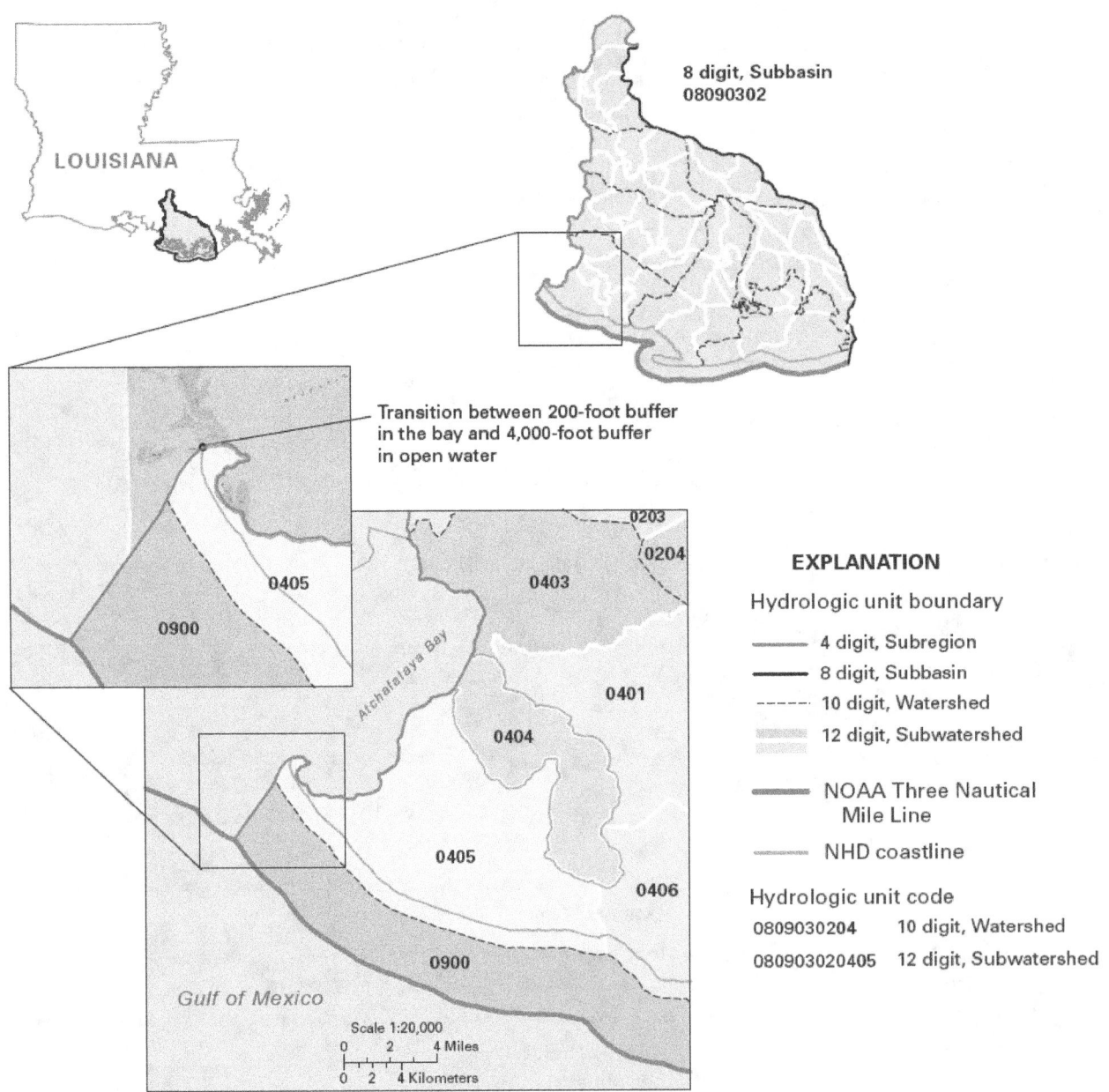

Figure 11.   A gradual transition is needed between two offshore depths or distances, such as the 200-foot buffer in this bay and the 4,000-foot buffer in this open-water area of offshore Louisiana. (NHD, National Hydrography Dataset; NOAA, National Oceanic and Atmospheric Administration)

3.6.2   Water and Frontal Hydrologic Units

Hydrologic units in coastal areas may drain either to a single outlet (standard hydrologic unit) or to multiple outlets (frontal drainage). Frontal hydrologic units are areas that drain to multiple points along a coast, and they can be separated by standard hydrologic units flowing into the waterbody. After the standard 10- and 12-digit hydrologic units are delineated within the designated size criteria (section 3.4), coastal and small stream drainage areas, often with radial or centripetal drainage, will remain where mainland or island coastal outlet areas are fragmented. These frontal hydrologic units are typically included with offshore hydrologic units. The outer extent of these coastal units is delineated at an offshore boundary, either the NOAA Three Nautical Mile Line or by criteria as determined by State partners. At a minimum, extend the boundaries of coastal hydrologic units to intersect the first offshore boundary encountered (fig. 12). Nearshore delineations must transition smoothly between physical features, and the edges must match at State boundaries. When coastal areas transition from bays, sounds, and estuaries to open ocean, the nearshore delineation can be continued but it is not required. If a frontal unit includes an 8-digit hydrologic unit boundary, then the delineation must be extended to all offshore boundaries, preferably along submerged ridges.

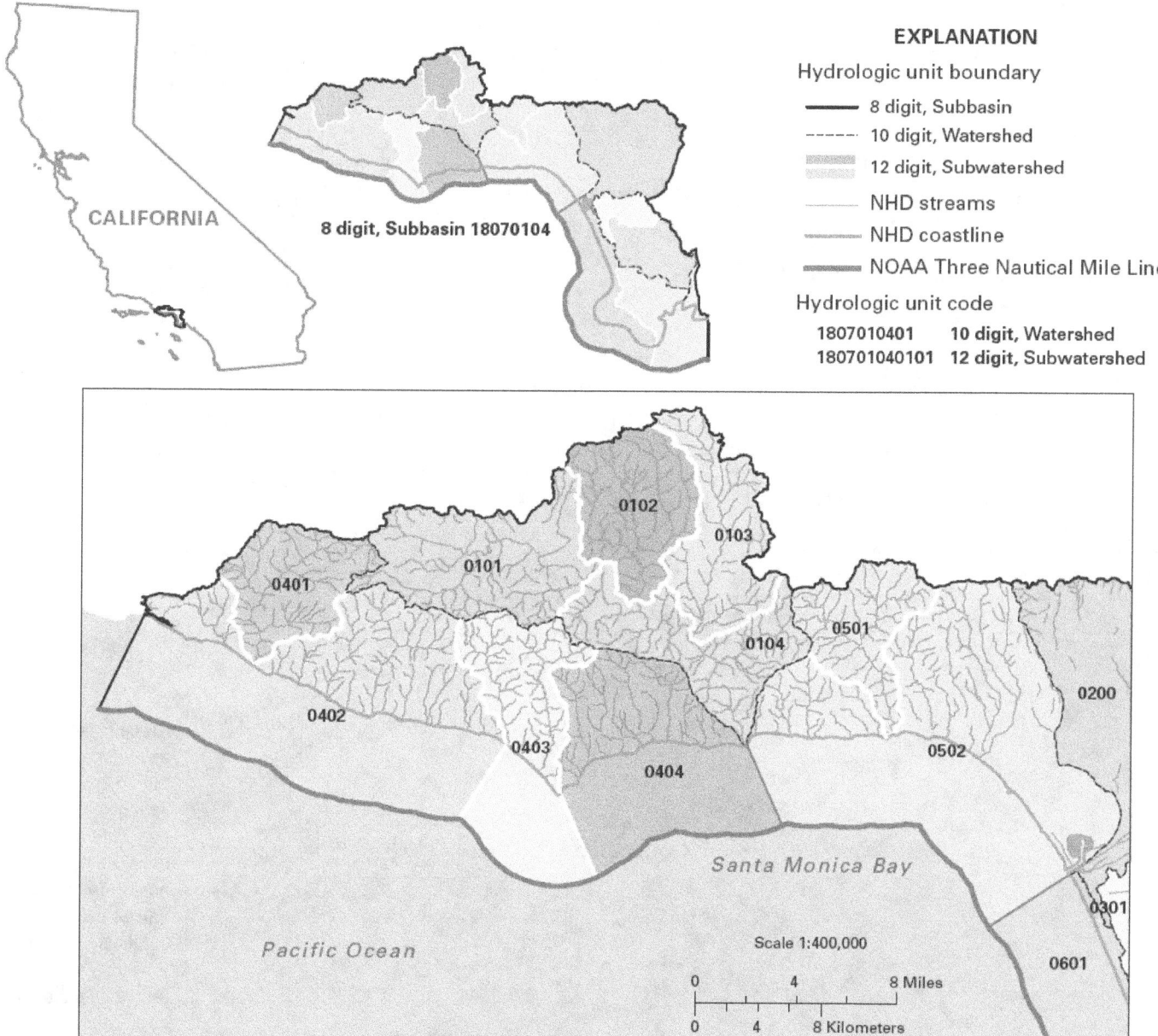

Figure 12.   The outer extent of these standard and frontal hydrologic units in the Santa Monica Bay area, California, is delineated to the first offshore boundary, the NOAA Three Nautical Mile Line. (NHD, National Hydrography Dataset; NOAA, National Oceanic and Atmospheric Administration)

3.6.3 Islands

Delineate island hydrologic units based on their size and proximity to each other and to adjacent land. An island large enough to be its own hydrologic unit can be delineated as such and drains into the surrounding water unit or ocean. The area of an island unit includes the land area and offshore extent. Subdivide island land area into hydrologic units consistent with unit size criteria. An island located between water hydrologic units may be bisected to indicate that surface water flows to the adjacent hydrologic units. This technique applies if a single barrier island or a string of islands along a shoal or reef functions as the hydrologic divide. The hydrologic-divide concept can be applied to any formation that provides a hydrologic barrier to flow, such as a peninsula or an isthmus.

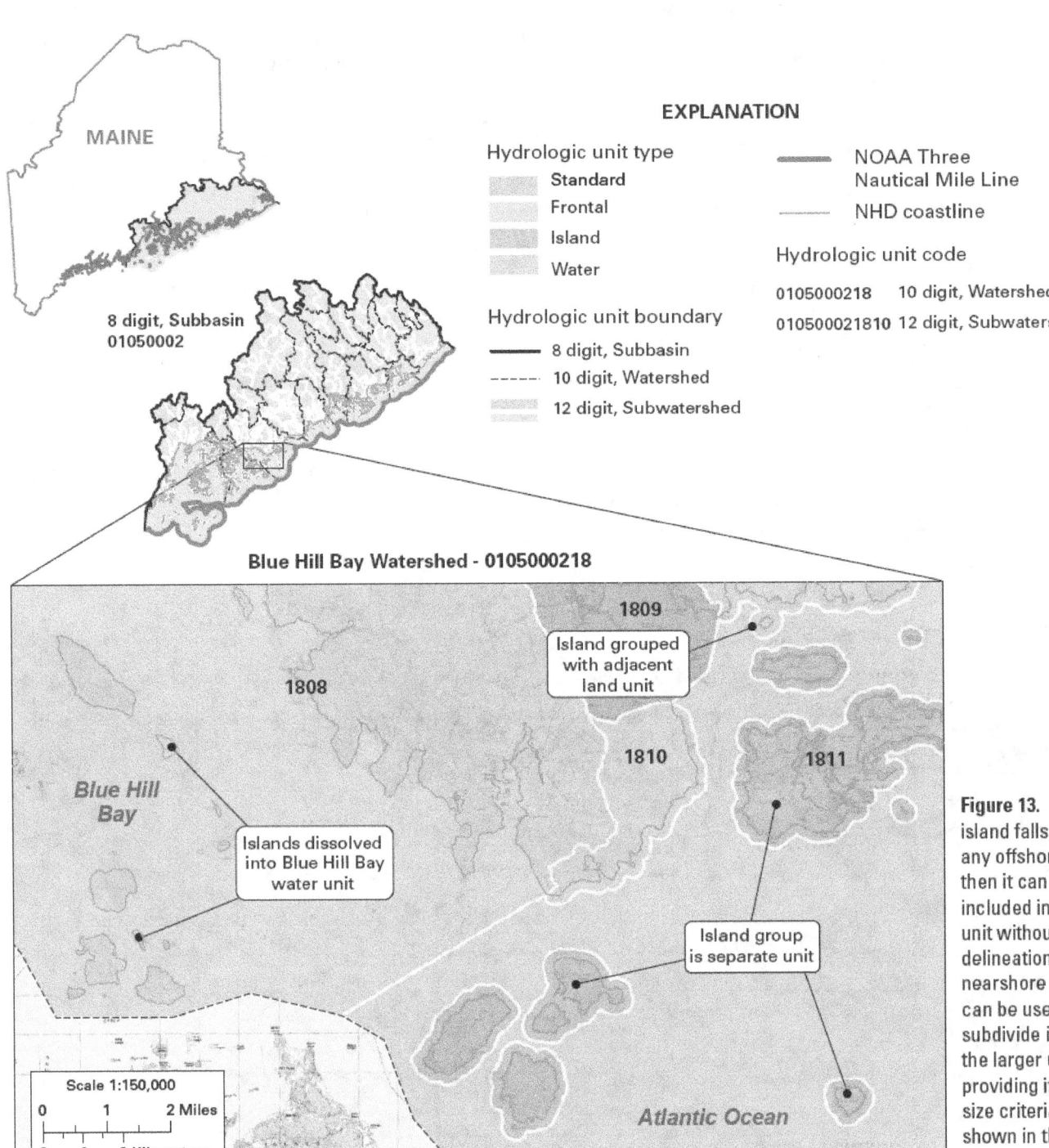

Figure 13. If an island falls within any offshore unit, then it can be included in that unit without further delineation or a nearshore buffer can be used to subdivide it from the larger unit providing it meets size criteria as shown in the Blue Hill Bay area in Maine.

An island too small to be its own hydrologic unit can be grouped with a nearby land unit or with another nearby island or group of islands, such as a barrier island. If an island falls within any offshore unit, then it can be included in that unit without further delineation or a nearshore buffer can be used to subdivide it from the larger unit providing it meets size criteria (fig. 13). If an island is beyond the mainland NOAA Three Nautical Mile Line, such as the Channel Islands off the California coast, then the outer extent of the island hydrologic unit is the NOAA Three Nautical Mile Line. The nearshore delineation is optional (fig. 14).

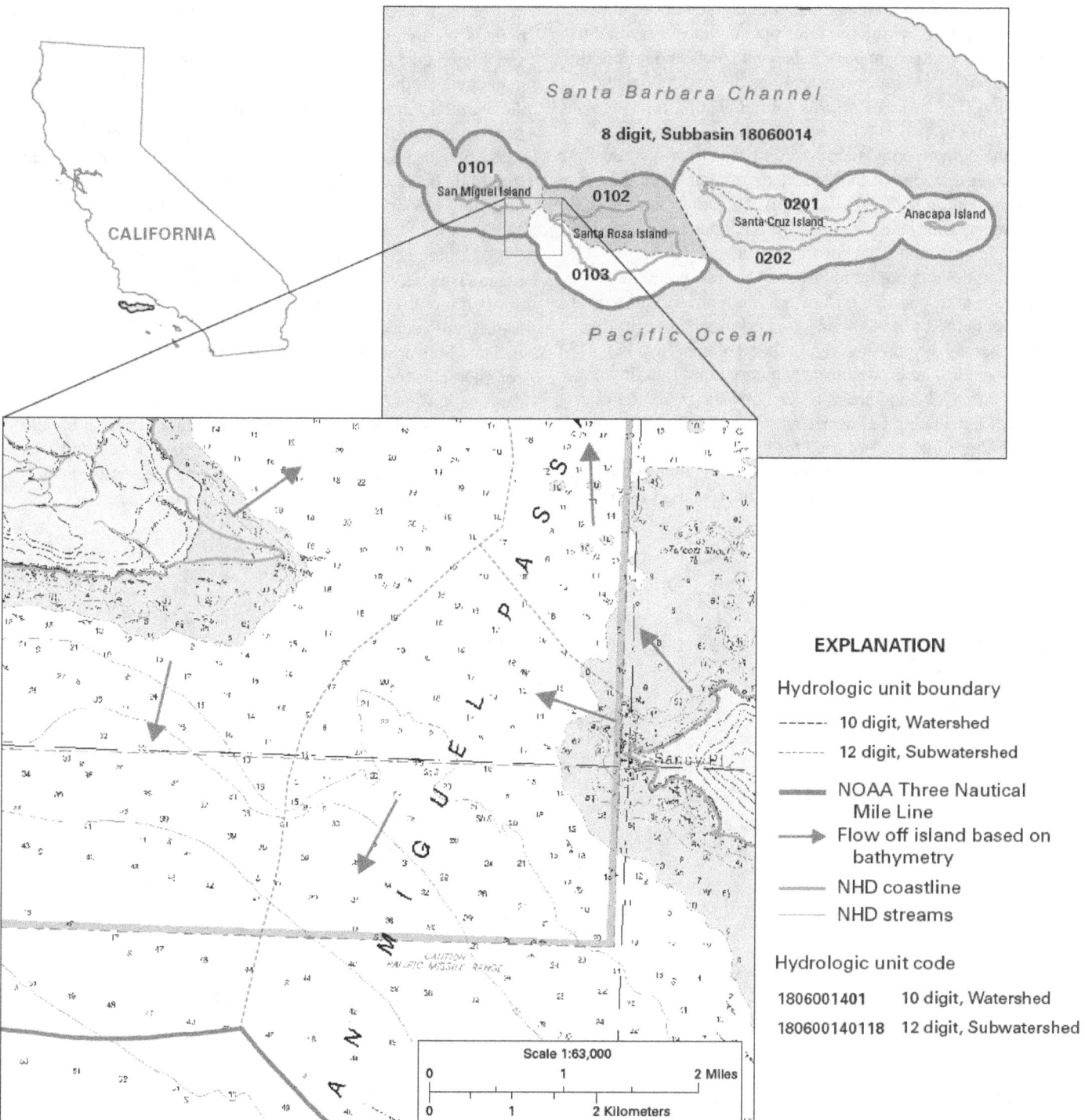

Figure 14. For an island that is beyond the mainland NOAA Three Nautical Mile Line, such as the Channel Islands off the California coast, the outer extent of the island hydrologic unit is the NOAA Three Nautical Mile Line. The nearshore delineation is optional.

3.6.4   Reservoirs and Natural Lakes

Delineation of hydrologic unit boundaries is often complicated by the presence of reservoirs or large natural lakes, because they can obscure the natural drainage system. The effect of the pool area and its underlying drainage pattern on the delineation of hydrologic units depends on the size of the pool area and the amount of fluctuation in the normal pool area. To delineate hydrologic units that include reservoirs, use depth data or historic maps of legacy channels and ignore the reservoir pool. Let underwater features direct the flow and delineation. The order of priority for delineating reservoirs and lakes follows.

If the uninterrupted natural drainage network would form a unit larger than one of the typical size at that level, then consider subdividing the hydrologic unit at a dam or natural outlet of the reservoir or lake. This decision will depend on adjoining hydrologic units and adjacent slope areas.

Avoid delineating boundaries to the reservoir's normal, average, or high pool. To ensure that all of a reservoir's pool area is accounted for, subdivide the hydrologic units to the legacy channel system underlying the pool area (fig. 15).

For natural lakes, delineate tributary areas that flow directly into a lake and are of a size consistent with the hydrologic unit level being delineated as standard hydrologic units. Because shorelines are not used for hydrologic unit boundaries, nonstandard areas that drain into a lake can be combined with the lake (fig. 16). If the resulting unit is larger than the size guidelines, then it can be subdivided based on bathymetry or it may be left intact if the subdivision is not defensible.

Permanent, large-scale water bodies, such as lakes and reservoirs having historically documented permanent pools, may require exceptions to these priorities. When large water bodies are adjacent to several 8-digit hydrologic units, hydrologic unit delineations without subdivision along legacy channels or bathymetry may be accepted into the WBD. Where the boundaries are established and recognized within State law and a legal lake elevation has been extensively used for water management, the hydrologic unit delineation may be based on the legal lake elevation. The interior of the lake is delineated as a water hydrologic unit. Large lakes or reservoirs delineated as hydrologic units, such as the Great Salt Lake in Utah, should be listed with their legal lake elevation in the detailed overview section of the metadata and identified with a Line Modification attribute value of SL for shoreline.

A unique coastal process was applied in the Great Lakes Basin. This treatment was developed by the WBD–NTC through months of coordination and collaboration with Canadian Federal and Provincial partners as well as all WBD In-State Stewards. This treatment is consistent with and complements the integration of the WBD and NHD with sister datasets in Canada, and it has been carefully developed to accommodate those datasets. WBD In-State Stewards should consult with the WBD–NTC prior to editing any coastal areas of the Great Lakes. Documents describing the evolution and agreed-upon method can be obtained through the WBD–NTC.

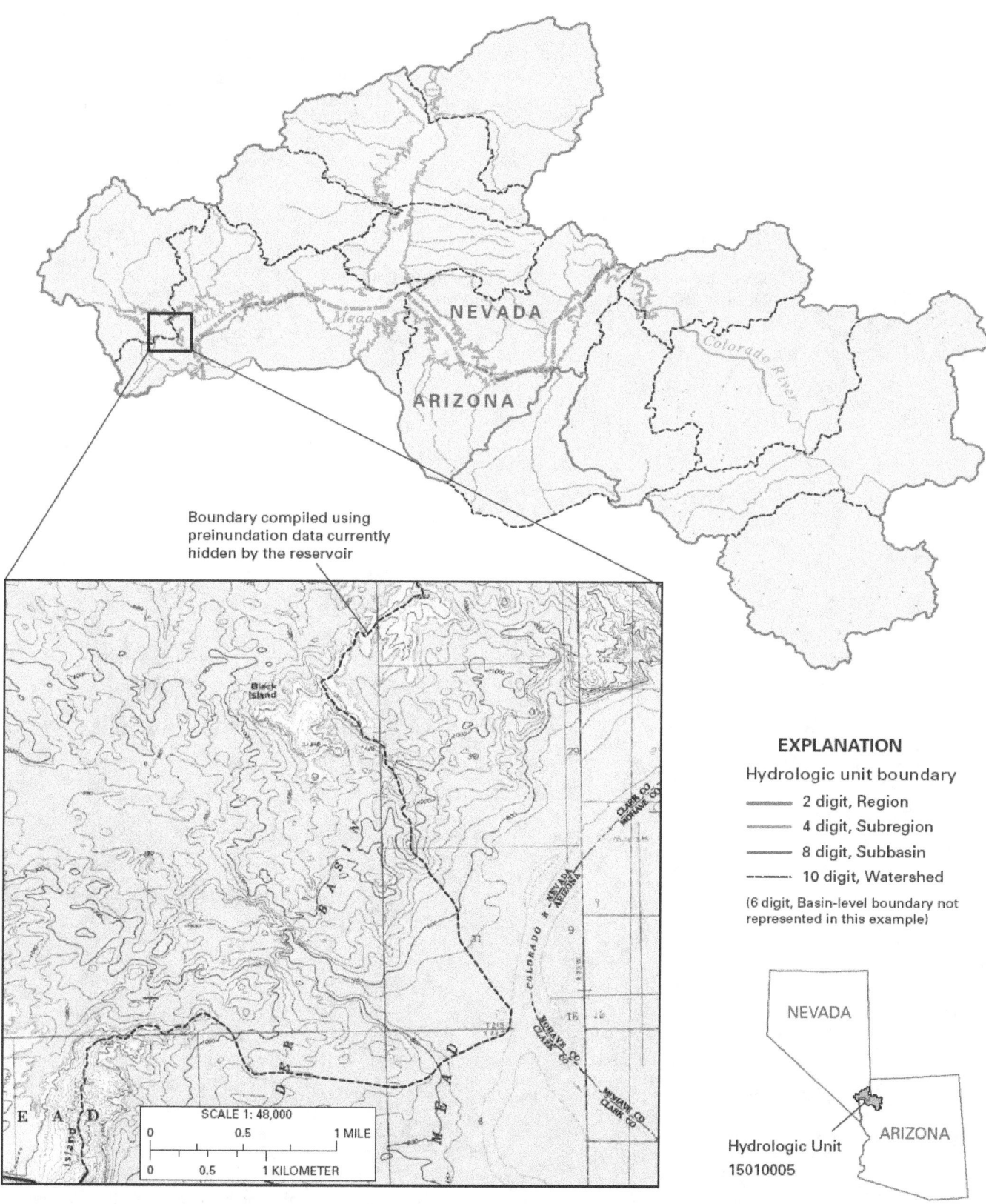

Figure 15. To delineate the hydrologic unit boundary within a hydrologic unit that contains a major reservoir, such as the Lake Mead area of Arizona and Nevada, the preferred treatment is to use a topographic map showing preinundation contours of underwater features if such a map is available.

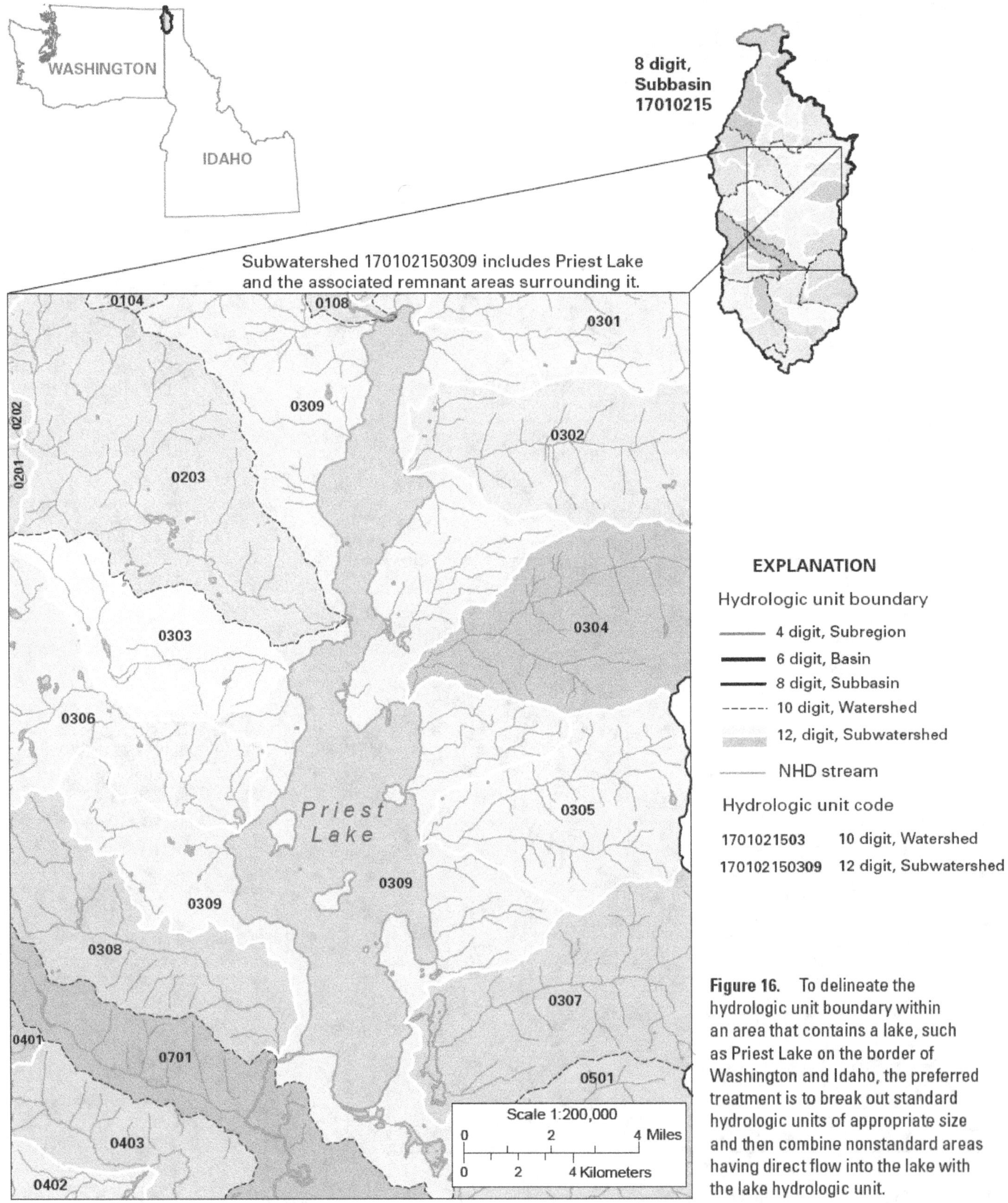

WASHINGTON

IDAHO

8 digit,
Subbasin
17010215

Subwatershed 170102150309 includes Priest Lake
and the associated remnant areas surrounding it.

0104   0108   0301

0202   0201   0309

0203   0302

0303   0304

0306

0305

*Priest
Lake*   0309

0309

0308

0401   0307

0701   0501

0403

0402

Scale 1:200,000

0   2   4 Miles

0   2   4 Kilometers

EXPLANATION

Hydrologic unit boundary

———— 4 digit, Subregion

———— 6 digit, Basin

———— 8 digit, Subbasin

------- 10 digit, Watershed

▓▓▓ 12, digit, Subwatershed

———— NHD stream

Hydrologic unit code

1701021503   10 digit, Watershed

170102150309   12 digit, Subwatershed

Figure 16.   To delineate the
hydrologic unit boundary within
an area that contains a lake, such
as Priest Lake on the border of
Washington and Idaho, the preferred
treatment is to break out standard
hydrologic units of appropriate size
and then combine nonstandard areas
having direct flow into the lake with
the lake hydrologic unit.

3.7   Optional 14- and 16-Digit Hydrologic Units

Development of 14- and 16-digit hydrologic units is optional, based on specific State needs, and should be undertaken only if the WBD In-State Steward and partner agencies agree that there is value in developing 14- and 16- digit hydrologic units for their area of interest. Delineations should be driven by the hydrology of the region and may vary by region.

3.7.1   Criteria for 14- and 16-Digit Hydrologic Units

Delineations of 14- and 16- digit hydrologic units follow many of the same criteria for delineations of 10- and 12-digit hydrologic units. The following minimum criteria should be used when delineating 14- and 16- digit hydrologic units.

- Legacy delineations should be inventoried, examined, and used where applicable.

- 14-digit hydrologic units are nested within an associated 12-digit hydrologic unit.

- 16-digit hydrologic units are nested within associated 14-digit hydrologic unit.

- If 14-digit hydrologic units are delineated, then at a minimum they must be completed for the entire 12-digit hydrologic unit and similarly for the 16-digit hydrologic units within a 14-digit hydrologic unit.

- At a minimum, 14- and 16-digit hydrologic units must be aligned with the 1:24,000 scale Digital Raster Graphic (DRG) contours, or with updated USGS Topographic Maps. 14- and 16-digit hydrologic units should be delineated with coordination from adjacent jurisdictions to create a seamless layer. Work with representatives in adjacent States to use a consistent approach in delineating 14- and 16-digit hydrologic units.

- Higher resolution or current base information should be used, if it is available, documented, and agreed upon by State partners.

- 14- and 16-digit hydrologic units should not follow administrative boundaries but should instead be based on topography.

Include metadata to provide information about documented methodology and describe the procedures used, base information, reference, coordinating partners, and all other pertinent information.

3.7.2   Delineation Options

Several options are suggested for delineation of 14- and 16-digit hydrologic units. Because some of the current delineations were created before any national standards were in place, legacy data created to fit the specific needs of the agencies that developed these hydrologic units and delineations are tied to the current delineations. Differing landscapes and environments require flexibility in delineating 14- and 16-digit hydrologic units. Three options for examining legacy delineations are described here, and all three options follow the guidelines stated in section 3.7.1. The first option is the "Open" option, where the 14- and or 16-digit hydrologic unit may be delineated in a sound manner in accordance with the area's hydrology—this option provides the greatest amount of flexibility. The second and third options are subsets of this open option. The second option is a "Standard and Remnant Area" option that breaks out standard 14- and/or 16-digit hydrologic units based on major tributaries along the main stem and has a remnant area left over. This option may be the most beneficial for project work usually done for the 14- and 16-digit hydrologic unit. The third option is the "Conventional" option, which uses the existing criteria stated for delineating 10- and 12–digit hydrologic units but uses smaller size criteria for the 14- and 16-digit hydrologic units. The third, conventional option would provide layers of consistent size and distribution that may be best for regional work, and it is the recommended option by the WBD–NTC because it fits most closely within the existing structure.

3.7.2.1   Open Option

The open option should be based on hydrology, but it does not have size and number criteria. The 14-digit hydrologic unit should nest within the 12-digit hydrologic unit. If any delineation is done within a 12-digit hydrologic unit, then all of the 14-digit hydrologic units should be completed within that 12-digit hydrologic unit. Because the 14- and 16-digit hydrologic units lack size restrictions, 14- and 16-digit hydrologic units could be the same size. For States that have delineated only 14-digit hydrologic units with size ranges from very small to quite large, but that still fit within a 12-digit hydrologic unit, the open option would fit into legacy delineations. This option would also work when a State has 12-digit hydrologic units that may have 14-digit hydrologic units delineated for just a few tributaries and the State has chosen to keep the rest of the 12-digit hydrologic unit as a large composite. Units are numbered in accordance with the "Federal Standards and Procedures for the National Watershed Boundary Dataset (WBD)."

The open option provides the most flexibility of the three options; however, it may be difficult for analysis over multiple hydrologic units. The open option also creates the least workload for stewardship in the short term.

3.7.2.2   Standard and Remnant Area Option

The standard and remnant area option examines major tributaries, breaking out standard hydrologic units along a main stem, and then classifies the remaining area as a remnant/composite area. Standard 14-digit hydrologic units that bound major tributaries along a 12-digit hydrologic unit main stem are developed. Hydrography and local knowledge are utilized to determine which streams are considered major within the 12-digit hydrologic unit, and the remaining areas are the remnant area 14-digit hydrologic units. Standard 14-digit hydrologic units might also be found in 12-digit remnant area, closed basin, frontal, island, or unclassified hydrologic units.

Standard 16-digit hydrologic units that bound major tributaries along a 14-digit hydrologic unit main stem are developed. Hydrography and local knowledge are utilized to determine which streams are considered major, and remaining areas are remnant area 16-digit hydrologic units. Standard 16-digit hydrologic units might also be found in remnant area 14-digit hydrologic units.

There are no size criteria for these 14- and 16-digit hydrologic units. Standard 14- and 16-digit hydrologic unit sizes are based on the size of identified tributaries. Remnant area 14- and 16-digit hydrologic unit sizes are based on the remaining areas after standard unit sizes have been delineated. There are no minimum or maximum numbers of 14-digit hydrologic units within a 12-digit hydrologic unit, and there are no minimum or maximum numbers of 16-digit hydrologic units within a 14-digit hydrologic unit.

Units are numbered in accordance with the "Federal Standards and Procedures for the National Watershed Boundary Dataset (WBD)."

3.7.2.3   Conventional Option

The conventional option uses the same criteria used for the delineation of 10- and 12–digit hydrologic units, but it alters the size limits to meet the 14- and 16-digit criteria. The 14-digit hydrologic units should typically be from 3,000 to 10,000 acres in size and should nest within the 12-digit hydrologic unit. The 16-digit hydrologic units should typically be from 100 to 3,000 acres in size and should nest within the 14-digit hydrologic units.

Units are numbered in accordance with the "Federal Standards and Procedures for the National Watershed Boundary Dataset (WBD)."

3.7.3   Review of 14- and 16-Digit Hydrologic Units and Inclusion in the WBD

Completed 14- and 16-digit hydrologic units are submitted under the same protocol as other WBD edits through the WBD In-State Steward. Metadata must accompany all 14- and 16-digit hydrologic units submitted for review.

The WBD In-State Steward submits the State-approved 14- and 16-digit hydrologic units and metadata to the WBD–NTC for technical review and for final quality assurance; for example, the WBD–NTC will check to ensure that units nest correctly without overlaps or gaps and that all required attribute fields are complete. If the units pass this review, then they can be added into the WBD. At this time (2012), 14- and 16-digit hydrologic units will not receive official certification.

4. Required Geographic Data Sources and Recommended Techniques for Boundary Delineation

There are many techniques for delineating hydrologic unit boundaries. In the past, delineation began with interpreting the hypsographic and topographic information found on 1:24,000-scale paper or polyester film source materials and manually drawing boundary lines based on the interpretation of those materials. These manually drawn lines were subsequently digitized using scanning and vectorization or tablet digitizing.

Methods that are more current employ GIS software to combine delineation and digitization of hydrologic unit boundaries on a computer display. Instead of interpreting the hypsographic and topographic information found on paper or polyester film sources as used in the previous technique, digital sources are used as interpretive, background images for delineation. The boundary lines are digitized on the computer screen using functionality typically found in GIS software. Digital background sources include 1:24,000-scale Digital Raster Graphics (DRG), updated USGS Topographic Maps, and orthoimagery and orthophotographs such as Digital Orthophoto Quadrangles (DOQ) or DOQQs.

GIS software has functions for deriving drainage areas from elevation data, such as the National Elevation Dataset (NED) or light detection and ranging (LiDAR) and supplemental hydrographic and physiographic data. The intermediate boundaries are then verified and refined using on-screen digitizing to NSSDA. WBD tools have been developed by the USGS for editing.

4.1 Map Scale and Map Accuracy

Delineations need to meet a 1:24,000 scale in the United States, except for Alaska at 1:63,360 scale, and 1:25,000 scale in the Caribbean, described in the NSSDA (Federal Geographic Data Committee, 1998b), at a minimum. This can be accomplished by using recommended hardcopy maps, digital geographic data, or combinations. Hardcopy sources include current USGS 1:24,000-scale topographic quadrangles and higher resolution (larger scale) local hardcopy maps. If higher resolution local maps or other project-acquired base sources are used to generate the hydrologic unit boundaries, they must be provided to the WBD–NTC to use for verification and review. Digital data sources include at a minimum the NRCS County Mosaic DRGs or updated USGS Topographic Maps at a 1:24,000 scale in the United States, except for Alaska at 1:63,360 scale, and 1:25,000 scale in the Caribbean, or larger scale DOQQs, and similar high-resolution digital data showing topography and surface-water features. Combinations of these geographic data should be used to delineate accurate, current hydrologic unit boundaries and to interpret areas of complex drainage regimes or geomorphologies.

Digital elevation data at least equivalent in scale to the USGS 30-meter, Level 2 Digital Elevation Models (DEMs) or the NEDs (both 30 meter and 10 meter) are acceptable for the delineation of draft hydrologic unit boundaries. All derivative draft or raster-based delineations will require smoothing and a thorough verification and adjustments to 1:24,000-scale or better topographic data such as the NRCS 1:24,000-scale County Mosaic DRGs or updated USGS Topographic Maps. In Alaska and the Caribbean, USGS topographic maps (in digital and hardcopy formats) at 1:63,360 scale or 1:25,000 scale, respectively, may be used.

The 8-, 10- and 12-digit hydrologic units must be delineated from and georeferenced to a minimum horizontal accuracy of 1:24,000 scale in the United States, except for Alaska at 1:63,360 scale and 1:25,000 scale in the Caribbean, to meet the *National Standard for Spatial Data Accuracy* (Federal Geographic Data Committee, 1998b). For example, to quantify 1:24,000-scale horizontal accuracy as it applies to the delineation of WBD-compliant hydrologic units, a hydrologic unit boundary must fall within a buffer of 40 feet or 12.2 meters of a well-defined point on a 1:24,000-scale topographic map. Geospatial positioning accuracy standards are defined and stated in documents by the Federal Geographic Data Committee (1998b).

4.2 Source Maps

This section describes the types of hardcopy and digital-format source maps that can be used to delineate hydrologic units. The sources described are not an exhaustive list. Delineation of some hydrologic unit boundaries may require use of larger scale products or newly developed products. Each source map or source data product's scale and series must be recorded in the metadata (section 6.2.4) and in the LineSource field (section 6.2.2.6.3).

4.2.1 Base Maps

Use printed USGS 1:24,000-scale topographic quadrangles for hardcopy delineations and NRCS 1:24,000-scale County Mosaic DRGs, or updated USGS Topographic Maps for compiling hydrologic unit boundaries on a computer display. USGS topographic quadrangles can be obtained at *http://www.usgs.gov/pubprod/maps.html*, and NRCS County Mosaic DRGs can be obtained at *http://datagateway.nrcs.usda.gov/*. Blueprints or similar facsimiles of the USGS 1:24,000-scale topographic maps can be distorted and should not be used. In areas where 1:24,000-scale base maps are not available, the USGS 1:25,000-scale, 1:63.360-scale, and/or USGS 1:100,000-scale maps may be used to generate lines. The NHD should always be referenced when generating hydrologic unit boundaries. Other data sources and map products can be used with the digital, 1:24,000-scale DRGs to facilitate the interpretation of hydrologic unit boundaries. Documentation of all supplemental base and source maps, digital and hardcopy, should be recorded in the LineSource field (section 6.2.2.6.3) as well as in the general metadata (section 6.2.4).

4.2.2 Hydrologic Unit Maps

Historically, smaller scale hydrologic unit maps provided a framework for hydrologic boundary information. These included the USGS State Hydrologic Unit Maps at 1:500,000 scale, National Atlas Hydrologic Unit Boundary data at 1:2,000,000 scale, or the Hydrologic Units of the United States data at *1:250,000 scale*. These and other hydrologic unit maps produced by NRCS, USDA–FS, USGS, and State and local entities were used to determine the general location and level of complexity of 10- and 12-digit hydrologic units. The completed WBD supersedes all previously used information as the hydrologic unit framework for the Nation.

4.2.3 Reference Maps

The following reference maps may be useful for delineating hydrologic unit boundaries where 1:24,000-scale base maps do not include sufficient detail for determining flow patterns based on topography. These supplemental reference layers or maps are useful for (1) documenting artificial flow delineations based on permanent features on the landscape, and (2) determining gravity flow in areas of extreme low relief or complex geomorphologies. In areas of flat terrain, interpolation between contours may be improved by reference to trails, old roads, or firebreaks in forested areas, all of which frequently follow drainage divides. The following types of digital and hardcopy maps may be used:

- county drainage maps;

- "as-built" plans, including diversions and ditches;

- flow-direction maps;

- NOAA nautical charts;

- ditch-canal maps;

- land-cover maps;

- soil-survey maps;

- orthophotos or other aerial photographs;

- major land-area resource maps;

- local highway or street profiles;

- local watershed-project maps.

When open-ocean delineation is required, use the NOAA Three Nautical Mile Line, as published on a NOAA nautical chart. Charts can be accessed at *http://www.nauticalcharts. noaa.gov/csdl/mbound.htm*. Retain as a permanent record all maps, measurement data, and other supplemental reference data/maps used in delineations (section 6.2.2.6.3).

4.2.4 Digital Data

Where topographic products are used, the NRCS 1:24,000-scale County Mosaic DRGs on-screen digital reference layer for delineating and digitizing hydrologic unit boundaries, or updated USGS Topographic Maps, are preferred. County Mosaics at 1:24,000 scale are available at *http://datagateway.nrcs.usda.gov*. Where these data are available, their use in delineation is acceptable.

The 30-meter and 10-meter NED DEMs are the preferred reference layers for using spatial modeling techniques to generate draft hydrologic unit boundaries. These DEMs can also be used with the County Mosaic DRGs and the NHD, as well as with local reference layers, to clarify challenging delineations. Both 30- and 10-meter NED data can be obtained from *http://ned.usgs.gov/*.

The 1:24,000-scale or better resolution NHD layers are the preferred hydrographic data layers to use with the NRCS 1:24,000-scale County Mosaic DRGs to determine best placement of hydrologic unit boundaries, to verify the hydrologic unit connectivity, and to aid in the naming of hydrologic units. The NHD layers are available at *http://nhd.usgs.gov/*.

Interferometric synthetic aperture radar (IfSAR) and LiDAR products of 1:24,000 scale or better are also useful for determining the alignment of hydrologic unit boundaries, especially in areas of low relief. The use of these products for improvement of hydrologic unit boundaries is increasing as they become more widely available and affordable (section 4.3.4).

4.3 Hydrologic Unit Mapping Techniques

Drainage divides are usually determined by bisecting ridges, saddles, and contour lines of equal elevation. Hydrologic unit boundaries follow the middle of the highest ground elevation or the halfway point between contour lines of equal elevation (fig. 2). Where a tributary intersects the bank of a receiving stream, the hydrologic unit boundary should cross the tributary outlet parallel to the receiving stream channel. The hydrologic unit has only one outlet point, except in the case of deltas, braided stream networks, and coastal and lakefront areas (section 3.6). Hydrologic unit boundaries cannot be streams (section 3.2).

Manual, digital, and semiautomated methods can be used to generate hydrologic unit boundaries. Procedures for completing delineation, mapping, and digitizing differ among these options, but all can produce boundaries that meet the 1:24,000 scale in the United States, except for Alaska at 1:63,360 scale and 1:25,000 scale in the Caribbean. Delineating boundaries in areas of complex or flat terrain or complex hydrography requires careful attention to scale and source data for accurate interpretation. Sections 3.3 and 3.4 give the general size criteria for 10- and 12-digit hydrologic units and the recommended distribution of those units within an 8-digit hydrologic unit.

4.3.1 Manual Techniques for Delineating Hydrologic Unit Boundaries

When delineating hydrologic units by drawing boundary lines on hardcopy maps, use, at a minimum, USGS 1:24,000-scale topographic map contours, elevations, and drainage patterns to interpret and delineate the hydrologic unit boundaries. Other supplemental geographic data such as county drainage maps, State hydrologic unit maps, and aerial photographs (section 4.2.3) may be used with the USGS 1:24,000-scale hardcopy topographic maps to facilitate interpretation of hydrologic units.

4.3.2 Digital Techniques for Delineating Hydrologic Unit Boundaries

When delineating and digitizing 1:24,000-scale hydrologic units on the computer screen, use the NRCS 1:24,000-scale County Mosaic DRG contours, or updated USGS Topographic Maps, elevations, drainage patterns, and the NHD at a minimum, to interpret and delineate the hydrologic unit boundaries. Be advised that shifts have been detected when base information from online mapping services is used. In special cases—for example, where topography is complex or flat—the scale of digitizing may decrease depending on the resolution of the base or source data. When data layers such as the NED, NHD, and DOQQs are used with the 1:24,000-scale DRGs to facilitate interpretation of hydrologic units (section 4.2.4), delineating on the computer display using GIS software may require digitizing 10- and 12-digit hydrologic units at a scale of approximately 1:7,500. Before delineation of an 8-digit hydrologic unit is complete, the perimeter boundary associated with the hydrologic unit must also be reviewed and updated to meet NSSDA 1:24,000-scale map accuracy standards.

4.3.3 Spatial Modeling Techniques for Delineating Hydrologic Unit Boundaries

Image processing, GIS, and hydrologic modeling applications can be used to manipulate DEM data, creating derivative data that represent landform features and drainage-network patterns. DEMs are available in various horizontal and vertical resolutions. Digital elevation data with resolutions at least equivalent to the vertical and horizontal resolutions of either the 30-meter, Level-2 DEMs or NED *(http://ned.usgs.gov/)* can be used to develop a draft or preliminary delineation of hydrologic units, which will require further refinement to meet NSSDA 1:24,000-scale map accuracy standards. DEMs

or NED are combined with other geospatial data, such as USGS NHD, using GIS to simulate drainage networks, stream courses, and direction of flow by applying hydrologic models. Maps generated from digital hydrography data with flow-direction arrows also are helpful for delineating hydrologic units.

Depending upon the spatial modeling technique used, the data and its consistency, the software applications used, and other characteristics, the DEMs and NED will most likely provide a generalized depiction of landforms and drainage networks. This is especially apparent in areas of moderate and low topographic relief or complex hydrography. For this reason, all DEM- and NED-generated boundaries should be independently checked against 1:24,000-scale DRGs, or updated USGS Topographic Maps, DOQs, 1:24,000-scale NHD, or larger scale data. In some situations, particularly in flat areas, elevation data can be wrong or misleading and the drainage network in the NHD needs to be considered. Likewise, drainage patterns portrayed in the NHD, in similar situations, can be wrong or misleading and need to be checked against elevation data. Adjustments to boundaries and confluences at all levels will be required to ensure a 1:24,000 scale in the United States, except for Alaska at 1:63,360 scale and 1:25,000 scale in the Caribbean. A detailed description of the source elevation model must be documented in the metadata.

A similar semiautomated production and review process can be followed for the 30-meter-pixel-size bathymetric DEMs for major estuaries and sounds available nationally from NOAA.

4.3.4 Delineation Using High-Resolution Base Products

New technology brings new opportunities for upgrading the hydrologic unit boundaries in the certified WBD with higher resolution versions derived from extremely detailed DEMs. The two preeminent high-resolution DEM types are called IfSAR and LiDAR. Updates to WBD using higher resolution base products are tracked on the existing boundaries using the LINESOURCE field (section 6.2.2.6.3).

Using high-resolution elevation data from a trusted, authoritative data source is important to the success of any project. WBD In-State Stewards should consult with the WBD–NTC prior to beginning delineations using LiDAR or IfSAR base products to obtain the most recent documentation about supplementary metadata requirements. Final linework generated from LiDAR or IfSAR products should be smoothed to eliminate excess vertices to be cartographically compatible with all other WBD linework.

4.4 Updating and Revising 2-, 4-, 6- and 8-Digit Hydrologic Units

Hydrologic unit names and codes are common identifiers used by many agencies for reporting hydrologic unit characteristics. Many reports are tied solely to the name or code of the hydrologic unit. During the creation of the WBD, it was stated that the longstanding, existing 2-, 4-, 6-, and 8-digit hydrologic units would not have any substantial boundaries, codes, or names changed as the larger scale product was developed from the 1:250,000 scale. Because of significant hydrologic inaccuracies, international border harmonization, or coastal delineations, some of these boundaries, codes, and names changed during the creation of the higher resolution WBD. These rare instances have been reviewed and officially accepted by the WBD–NTC and WBD Steering Committee as of January 2012.

In the future, names and codes should not be changed for the 2-, 4-, 6-, and 8-digit hydrologic units. Boundaries should not be altered except for minor adjustments that will improve the data, such as refinements made because of the use of higher resolution base information. For any potential alteration other than these minor adjustments, consult the WBD–NTC. For coastal delineations, characteristics that may include a submerged river basin may warrant boundary, code, or name changes from legacy information. These instances should be carefully coordinated with the WBD–NTC on a case-by-case basis.

The 8-digit hydrologic units have been used and referenced so extensively in water-resource activities nationwide that major changes to them should be made only in cases of delineation error or major landform changes due to natural phenomena or human activity. Some examples include the removal of a dam, flow changes caused by earthquakes, construction of new reservoirs, embankments, or levees, volcanic eruptions, massive landslides, or hurricane damage. The identification of errors in the original digitized work may also lead to an update for some or all levels within an 8-digit hydrologic unit. For example, major revisions include those that place entire stream reaches (not small pieces of headwater reaches) in different 8-digit hydrologic units or those that recode contiguous areas approximating or exceeding the size of 12-digit hydrologic units of 10,000 to 40,000 acres. Inclusion of new offshore areas for coastal hydrologic units and revisions or additions to current terrestrial hydrologic unit delineation resulting from offshore delineations are considered revisions.

If the locations of 8-digit hydrologic unit boundaries need to be changed to be correct with the hydrology of an area, then notify the national WBD–NTC through the WBD In-State Steward. Obtain assurance from the WBD–NTC that proposed changes are acceptable before revising 8-digit hydrologic unit boundaries. When hydrologic unit boundaries have major revisions, update the area measurements, and note revisions as "revised" when the new data are released. Keep a record of all of the changes to the 8-digit hydrologic unit boundaries.

5.   Coding and Naming 10-, 12-, 14- and 16-Digit Hydrologic Units

5.1   Hydrologic Unit Levels

Guidelines for delineating the optional subdivisions of 14- and 16-digit hydrologic units are available from the NRCS and the USGS. Delineation of 14- and 16-digit hydrologic units is not required; however, if a 14-digit unit is delineated, then subdivide that entire 12-digit hydrologic unit. If a16-digit hydrologic unit is delineated, then subdivide that entire 14-digit unit.

The eight different levels of hydrologic units and their characteristics are shown below. Figure 3 provides an example of this for the 2- through 12-digit hydrologic units.

Hydrologic unit name	Historical name	Average size (square miles)	Approximate number of hydrologic units
2 digit	Region	177,560	21 (actual)
4 digit	Subregion	16,800	222
6 digit	Basin	10,596	370
8 digit	Subbasin	700	2,270
10 digit	Watershed	227 (40,000–250,000 acres)	20,000
12 digit	Subwatershed	40 (10,000–40,000 acres)	100,000
14 digit	(None)	Open	Open
16 digit	(None)	Open	Open

5.2 Coding 10-, 12-, 14- and 16-Digit Hydrologic Units

This section provides guidance on coding 10- and 12-digit hydrologic units, and optional 14- and 16-digit hydrologic units, for WBD compliance, as well as their relation to the 1:250,000-scale coding. The 1:250,000-scale dataset does not contain a consistent coding structure. Avoid changing the existing numbering of the 8-digit codes unless there is a topographic or hydrologic justification to do so. If there is a legitimate reason to alter a location or code of an 8-digit hydrologic unit, then the WBD In-State Steward should notify the WBD–NTC before making revisions (section 4.4). Assign a new unique numbered code to each 10- and 12-digit hydrologic unit. Maintain the additional 2-digit field length for successive hydrologic units. Coordinate the coding within an 8-digit hydrologic unit across State boundaries.

Avoid complicating the coding of hydrologic units. Number the hydrologic units sequentially, beginning upstream at the hydrologic unit with the uppermost outlet, including along the mainstem, and proceed downstream. Downstream codes are in ascending order within each level (lower numbers always flow into higher numbers). All numbers from the lowest to the highest must be used. For example, one can start at the upstream end of the drainage and code the first 10-digit hydrologic unit as 0908020301, code the next 10-digit unit downstream as 0908020302, and so forth. No numbers are skipped. The main-stem hydrologic unit is assigned the highest number when outlets are adjacent, or break at the same place, as shown in figure 17 (12-digit hydrologic units 170601020104 and 170601020105).

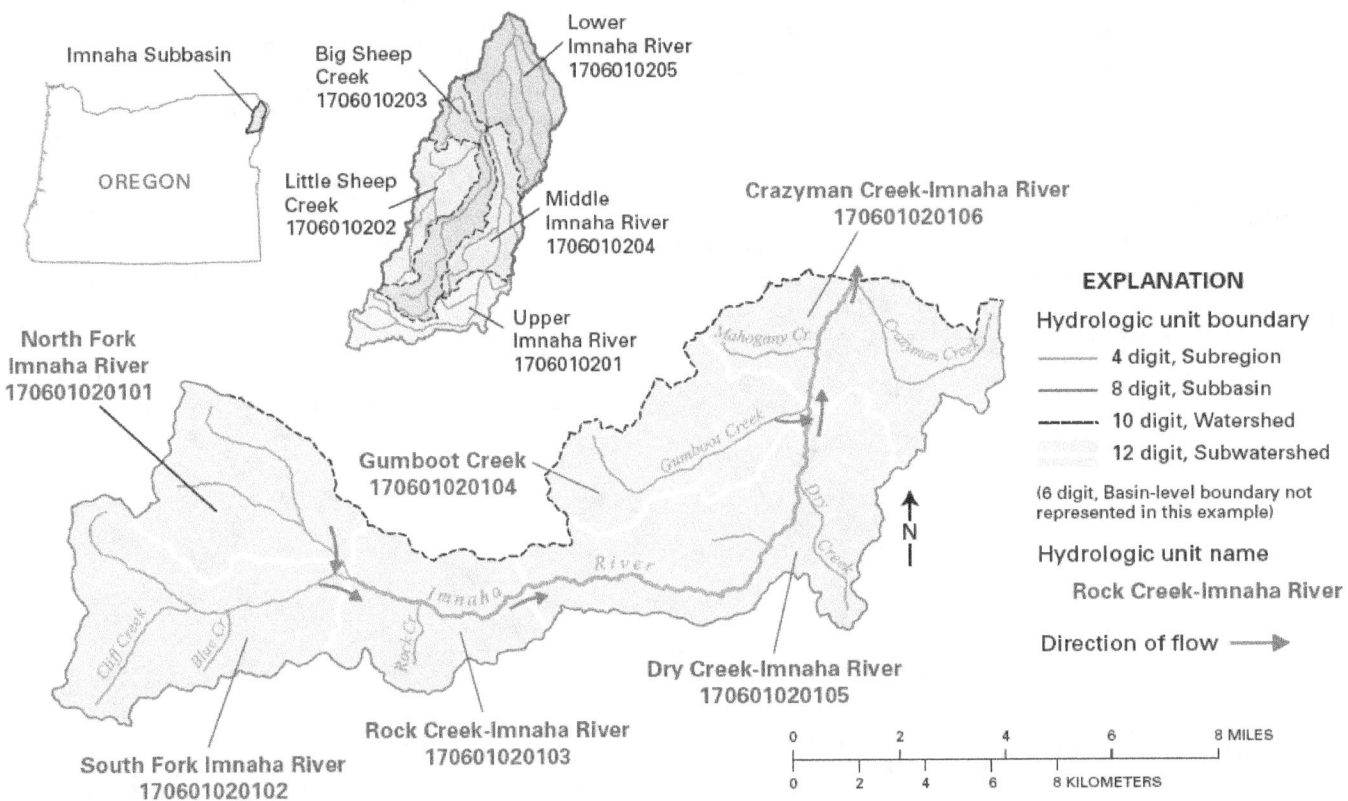

Figure 17. Each hydrologic unit should be coded sequentially based on the location of the outlet, starting with the uppermost stream outlet. The downstream code should always be a higher number than the upstream code for 10- and 12-digit hydrologic units. The main-stem hydrologic unit should carry the higher code when outlets are adjacent or break at the same place. Codes are shown for the Upper Imnaha River 10-digit hydrologic unit.

If there is no scientific justification for subdividing a hydrologic unit, then code the hydrologic unit 00. For example, if a 10-digit hydrologic unit (1020305108) cannot be subdivided on the basis of hydrologic principles, then the 12-digit hydrologic unit should be coded with "00" (for example, 102030511800). The "00" represents no subdivision from the next higher level.

An example of the numbering of hydrologic units:

Numbering sequence step	Field	Numbers	Level	Common name
1	2 digit	01	1	Region
2	4 digit	0108	2	Subregion
3	6 digit	010802	3	Basin
4	8 digit	01080204	4	Subbasin
5	10 digit	0108020401	5	Watershed
6	12 digit	010802040101	6	Subwatershed
7	14 digit	01080204010103	7	No historical name
8	16 digit	0108020401010302	8	No historical name

5.2.1 Coding Coastal Areas That Contain Frontal Units

This section provides coding rules for frontal hydrologic units. Frontal hydrologic units are areas that drain to multiple points along a coast, separated by standard watersheds flowing into the water hydrologic unit. Frontal hydrologic units that do not meet the size criteria using the combined area of the nearshore buffer and the frontal unit should be grouped together to make larger units. Frontals can also be grouped with nearshore or open-ocean hydrologic units. When aggregating these frontal hydrologic units, group those that are flowing into the same hydrologic feature, for example, a bay, lake, or estuary (see fig. 10, unit 1810).

Extend any boundaries that truncate onshore to offshore boundaries, based on underlying topography, provided this does not result in major alterations of the land codes. Code these coastal areas in a clockwise sequence from north or east, depending on the orientation of the hydrologic unit (fig. 10).

5.2.2 Coding Islands

Islands large enough to be subdivided can be coded in the same pattern for standard and frontal-unit coding, in a clockwise sequence from north or east, depending on the orientation of the island. Islands beyond the mainland NOAA Three Nautical Mile Line should use an island NOAA Three Nautical Mile buffer and be assigned a unique code. Islands can be grouped and need not be uniquely identified.

The coding of other islands depends on how the State chooses to delineate the outer extent of their offshore boundaries, which, along with proximity, determines island grouping.

An island can be coded individually when the combined area of the nearshore buffer and the island meets the size criteria. Island groups containing several islands can be coded individually in a clockwise sequence from north or east, depending on the orientation of the islands. Smaller islands, whose offshore boundaries do not intersect another hydrologic unit, can be grouped with the closest island or land hydrologic unit and given the same code and name as the larger unit (fig. 13).

5.3 Naming Protocol for 10-, 12-, 14- and 16-Digit Hydrologic Units

The numerical hydrologic unit code is the primary unique identifier for each hydrologic unit; however, the numerical identifier alone makes it difficult to relate a hydrologic unit to a geographic location. Therefore, the naming of 10- and 12-digit hydrologic units, and optional 14- and 16-digit units, adds local and geographic identity to the hydrologic unit and is helpful for understanding the geographic location of the hydrologic unit. Hydrologic units are usually named after prominent water features in an area; however, if no water features are in an area, then hydrologic units may be named after prominent physical features.

Identify each 10-, 12-, 14- and 16-digit hydrologic unit with a feature name from the area being subdivided. Maintain consistent structure when assigning a name to a unique hydrologic unit. Use the following guidelines for naming 10- and 12-digit hydrologic units and optional 14- and 16-digit hydrologic units. Do not change the names of the 2-, 4-, 6-, and 8-digit hydrologic units.

5.3.1   Sources of Names for 10-, 12-, 14-, and 16-Digit Hydrologic Units

The 10-, 12-, 14-, and 16-digit hydrologic units should be named using feature names officially recognized in the GNIS. Options are contained in the GNIS of the USGS at *http://geonames.usgs.gov/*, including names on USGS 1:24,000-, 1:25,000-, or 1:63,360-scale topographic quadrangles, or in the NHD at *http://nhd.usgs.gov*. Avoid naming the hydrologic units after roads, streamflow-gaging stations, and political or administrative units.

5.3.2   Feature Priority of Names for 10-, 12-, 14-, and 16-Digit Hydrologic Units

Assign each 10-, 12-, 14-, and 16-digit hydrologic unit a name from the GNIS in this order of priority, advancing to the next level if no features of a type exist or are named:

First–HYDROLOGIC FEATURE (Examples include rivers, lakes, dams, and falls)

   Second–GEOLOGIC FEATURE (Examples include canyons, mountains, buttes, and peaks)

      Third–TOWN OR CITY (Town of Name, City of Name)

         Fourth–OFFICIAL LOCAL NAME (Examples include cemeteries, municipal or international airports)

If no water or prominent physical feature name is assigned, then the hydrologic unit code is used as a placeholder.

Exceptions may be made to this priority order based on feature importance. For example, a canyon may be more prominent than a hydrologic feature, such as an upstream spring; this is typical in arid areas. Avoid using identical names within an 8-digit hydrologic unit. All 10-digit hydrologic unit names within an 8-digit unit must be unique to each other, and all 12-digit hydrologic unit names within an 8-digit unit must be unique to each other.

5.3.3   Structure of Names for 10-, 12-, 14- and 16-Digit Hydrologic Units

Populate the hydrologic unit name attribute field using the following structure.

- Use the name of the major water feature within the hydrologic unit, spelled out exactly as listed in the GNIS, usually the water feature at the outlet of the hydrologic unit. Stream names are preferred, but sloughs, lakes, reservoirs, dams, bays, inlets, harbors, coves, falls, and springs may be used when they are the most important feature; for example, "Crescent Lake," "Sequim Bay," and "Grays Harbor."

- The word "Frontal" is reserved for coastal and lake areas that include multiple, nonconvergent streams associated with frontal hydrologic units. Name the hydrologic unit for the major hydrologic feature within the hydrologic unit and use "Frontal" as a prefix to the name of the hydrologic feature into which the unit drains, for example, "Squirrel Creek-Frontal Chesapeake Bay."

- When the same primary water feature exists within equivalent-level hydrologic units or if a main-stem stream is subdivided into more than one hydrologic unit, use the following hyphenated naming structure to create unique names. Append the primary water feature name (Imnaha River) onto a secondary water feature such as a large tributary name (Rock Creek), for example, "Rock Creek-Imnaha River," or "Dry Creek-Imnaha River" (fig. 17).

- When a major stream is subdivided into three hydrologic units along the main-stem stream, use the words "Upper," "Middle," and "Lower" together, for example, "Upper Imnaha River," "Middle Imnaha River," "Lower Imnaha River" (fig. 17) Avoid using "Upper," "Middle," and "Lower" individually or "Middle" with only "Upper" or "Lower."

- "Headwaters" or" Outlet" may be used with or without "Upper," "Middle," or "Lower." If a main-stem stream is subdivided into more than five hydrologic units, then avoid this naming convention; use the standard hyphenated name.

- "Upper" and "Lower" or "Headwaters" and "Outlet" "may be used in pairs when a stream is subdivided into two hydrologic units.

- Do not use "Upper," "Middle," and "Lower" with the hyphenated naming structure.

- The words "Headwaters" and "Outlet" may be used with the hyphenated naming structure or the "Upper," "Middle," and "Lower" naming structure.

- When bathymetry is used to delineate submerged morphologic features, the resulting coastal water unit may not contain any other named features except the main water body name (for example, Atlantic Ocean). For these units, use the "hydrologic unit code-primary water feature" for the name, for example, 030102051406-Atlantic Ocean.

- Islands should be named according to the following conventions.

  - When islands are large enough to be subdivided, use the standard naming convention for standard and frontal hydrologic units.

  - Islands large enough to be their own unit use the island name.

  - Hydrologic units composed of a group of islands use the name of the major island within that group.

5.4   International Borders

Delineating or coding across international boundaries should be coordinated with the WBD–NTC, which will then provide the forum for coordinating further at the Federal, State, and Provincial level. Delineations or attributes along international borders may require future adjustment.

5.4.1   United States Data Harmonization with Canada

The USGS, in coordination with the International Joint Commission, Natural Resources Canada, Agriculture and Agri-Food Canada, Environment Canada, and many State and Provincial partners, has completed the 8-digit hydrologic unit delineations across the United States-Canada border. Completion of the 10- and 12-digit hydrologic unit delineations is estimated to continue through 2014.

Within the WBD, 12-digit hydrologic unit resides the international unique code, and for Canada this is the BinCNUS polygon attribute field. The coding is unique to this border. The code is populated within the 12-digits for every shared 8-digit hydrologic unit along the border. The first three digits of the seven-digit number for the BinCNUS code represent the first harmonized level with Canada, which is at the United States 8-digit hydrologic unit level. The following two digits represent the 10-digit subdivision, and the subsequent two digits represent the 12-digit hydrologic unit level. These are merely sequential codes at each level, beginning in the north in Alaska, then south to Washington, and then east to Maine. Within the 8-digit hydrologic units, the numbering is sequential from headwaters to outlet for the next two nested levels.

5.4.2   United States Data Harmonization with Mexico

To facilitate hydrologic analysis applications, the USGS and the Mexican Instituto Nacional de Estadística y Geografía (INEGI) collaborated to harmonize shared hydrologic drainage areas along the United States-Mexico border and build a connected hydrographic network of surface-water features for the binational region. The harmonization process included participatory meetings between team leaders for the United States WBD and NHD and Mexico's Red Hidrográfica and Cuencas Oficiales to agree on standards, content, connectivity, and binational "crosswalk" attribution tables from each country. These data are available within the WBD.

Within the WBD, 12-digit hydrologic unit resides the international unique code, and for Mexico, this is the BinMXUS polygon attribute field. The coding is unique for this border. The code is populated within the 12-digits for every shared 8-digit unit along the border. The first two digits of the six-digit number for the BinMXUS code represent the first harmonized level with Mexico, which is at the United States 8-digit hydrologic unit level. The following two digits represent the 10-digit subdivision, and the subsequent two digits represent the 12-digit hydrologic unit level. These are merely sequential codes at each level, beginning in the west in California and running east to the tip of Texas. Within the 8-digit units, the numbering is sequential from headwaters to outlet for the next two nested levels.

6.   Geospatial Data Structure

6.1   Specifications for Geospatial Data Structure

The structure of the WBD is based on well-known GIS concepts for modeling geographic features as vectors. Lines are used to model the linear features of hydrologic units, whereas polygons are used to model the areal features. For both linear and areal features, associated attribution is stored in tabular format. As part of the integration with the NHD, the WBD geospatial data model has been updated. The model was developed under the USGS common vector model practices. These standards have been implemented across the USGS to insure that all data maintained and delivered by the USGS is consistent in structure, naming conventions, and metadata. The common vector model also seeks to avoid duplication of data across the multiple datasets delivered by the USGS.

6.1.1   Hydrologic Unit Geometry

Hydrologic unit boundaries are delineated and digitized as lines. The digital lines are processed using GIS software to form polygons. Recommended techniques for data capture are described in section 4. All typical geometric problems should be resolved, including dangling lines, sliver polygons, and missing or duplicate labels. Note that some of these issues are specific to a particular spatial data format, such as Environmental Systems Research Institute (Esri) Coverage format, and do not necessarily apply to other formats such as Esri Geodatabase format or Esri Shapefile format.

6.2   Database Schema for Attributing Hydrologic Unit Polygons

Each hydrologic unit polygon has associated attributes stored in tabular format. An attribute is defined by name, type, size, and valid values (or range of valid values). A collection of attributes is called a schema. There are two categories of schema: logical and physical. The logical schema refers in general terms to the information associated with hydrologic unit polygons; for example, every hydrologic unit polygon has a unique code. The physical schema, on the other hand, defines the specific format of the table used to store attribution; for example, one of the codes for a hydrologic unit polygon has the literal name, "HUC8," and is required to have eight characters, a value is required (cannot be left blank), and if needed, preceding zeros fill out all the characters. The physical schema is immutable; that is, attribution must adhere to literal field names, field types, and the order of fields in the table.

6.2.1   Changes to the Data Model Due to Integration with the NHD

Prior to integration of WBD with the NHD, polygons and lines were edited within the WBD. Under the new model, base classes are edited, and then derived classes are created from these edits. Base classes include attributes, lines, and points. Derived classes are the attributed polygons, organized by hydrologic unit digit code. An example of a derived class would be a polygon layer consisting of all of the 12-digit hydrologic units for the Nation. The derived classes are updated whenever edits are submitted to the WBD from the base classes.

6.2.2   Base Classes

Base classes include four feature classes and three tables. The feature classes include WBDLine, WBDPoint, and NWISLine. The tables include WBDAttributes, WBDNav and WBDFeatureTo HUMod. Base classes share five common attributes: PermanentIdentifier, SourceFeatureID, MetaSourceID, SourceDataDesc, and SourceOriginator. Some base classes are automatically created and assigned, and others must be populated by the WBD In-State Steward.

6.2.2.1   PermanentIdentifier

PermanentIdentifier is a unique 36-character field that identifies each element in the database exclusively. PermanentIdentifier is an automatically assigned code that always stays with each element. When an element is updated or changed, PermanentIdentifier links the element to the metadata record, documenting that change. PermanentIdentifier is also used to maintain relationship classes in the normalized data model. When an element is deleted or split, PermanentIdentifier stays with the original element and is not used again. In the case of a split, new permanent identifiers are assigned to the resultant parts.

Field Name	PermanentIdentifier
Field Type	Character
Field Width	40
Domain	Randomly generated alpha-numeric character
Value required?	Yes
Example	{5DD21DC6-3692-4197-889B-49E652AA43D0}

6.2.2.2 SourceFeatureID

Like PermanentIdentifier, SourceFeatureID is a long, unique code. This code identifies the parent of the feature if the feature is the result of a split or merge, and it is automatically generated and assigned.

Field Name	SourceFeatureID
Field Type	Character
Field Width	40
Domain	Randomly generated alpha-numeric character
Value required?	No
Example	{5DD21DC6-3692-4197-889B-49E652AA43D0}

6.2.2.3 MetaSourceID

MetaSourceID is another unique identifier that links the element to the metadata tables. This ID is generated and assigned automatically by the database and remains with the object permanently.

Field Name	MetaSourceID
Field Type	Character
Field Width	40
Domain	Randomly generated alpha-numeric character
Value required?	Yes
Example	{5DD21DC6-3692-4197-889B-49E652AA43D0}

6.2.2.4 SourceDataDesc

SourceDataDesc is a space provided for a brief description of the type of base data used to update or change the current WBD. The WBD In-State Steward completes this field as part of the metadata form.

Field Name	SourceDataDesc
Field Type	Character
Field Width	100
Domain	User-entered text limited to 100 characters
Value required?	Yes
Example	Montgomery County 1-meter LiDAR

6.2.2.5 SourceOriginator

SourceOriginator is the description of the agency that created the base data used to improve the WBD. The WBD In-State Steward completes this field as part of the metadata form.

Field Name	SourceOriginator
Field Type	Character
Field Width	130
Domain	User-entered text limited to 130 characters
Value required?	Yes
Example	USDA–FS LiDAR

6.2.2.6   WBDLine

WBDLine is the primary line representation of the WBD. This feature data class represents the official geometry of the WBD, and all polygon classes are derived from WBDLine. WBDLine is also the primary data class for geometry editing by WBD In-State Stewards. WBDLine is edited by changing the line geometry and attributes. The edits are then propagated throughout the other WBD-derived feature classes. WBDLine includes five common fields and four unique fields described below. WBDLine contains PermanentIdentifier, SourceFeatureID, MetaSourceID, SourceDataDesc, SourceOriginator fields. These common fields are automatically created and assigned. See the previous descriptions of these fields (sections 6.2.2.1–6.2.2.5).

6.2.2.6.1   Common Fields

WBDLine contains PermanentIdentifier, SourceFeatureID, MetaSourceID, SourceDataDesc, SourceOriginator fields. These common fields are automatically created and assigned. See the previous descriptions of these fields (sections 6.2.2.1–6.2.2.5)

6.2.2.6.2   WBDLine: HULevel

HULevel is a domain-based field, completed by the WBD In-State Steward, which indicates the WBD line level by use of a single digit. For example, a 12-digit hydrologic unit line would be a level 6 line, and a 4-digit hydrologic unit line would be a level 2 line. This field allows for cross-walking linked reports and external data that used the level classification.

The hydrologic unit level attribute indicates the relative position of each boundary line segment within the hydrologic unit hierarchy. Populate this field with the highest hydrologic unit level (smallest number) for each line as indicated by the record. A level is represented by numbers 0 through 8, with 0 at the top and 8 at the bottom of the hierarchy. For example, if a line segment is the boundary of a 2-digit hydrologic unit, then the value is 1. If a line segment is the boundary of an 8-digit hydrologic unit, then the value is 4, even though it is also a 10- and 12-digit hydrologic boundary. If a line segment ends at an international border and there is no information to complete the hydrologic unit beyond the United States boundary, then the value is 0. Where the NOAA Three Nautical Mile Line designation represents the outer extent of the hydrologic unit, the value is 1. Use one of the levels in the list below.

Level	Name (number of digits)	Common name
0	0	No data (international border)
1	2	Region
2	4	Subregion
3	6	Basin
4	8	Subbasin
5	10	Watershed
6	12	Subwatershed
7	14	No historical name
8	16	No historical name

Field Name	HULevel
Field Type	Long Integer
Field Width	6
Domain	1, 2, 3, 4, 5, 6
Value required?	Yes
Example	6

6.2.2.6.3   WBDLine: HUClass

HUClass is a domain-based field that indicates the minimum number of digits used to represent the hydrologic unit bounded by the line. The WBD In-State Steward completes HUClass entries.

Field Name	HUClass
Field Type	Long Integer
Field Width	16
Domain	2, 4, 6, 8, 10, 12, 14, 16
Value required?	Yes
Example	12 (Subwatershed)

6.2.2.6.4 WBDLine: LineSource

LineSource represents the code for the base data used for delineating hydrologic unit boundaries. There is an agreed upon standard for the abbreviation of the line source information, which indicates if the line was created using USGS 1:24,000 DRG's, a DEM, LiDAR, or some other base data source. The WBD In-State Steward completes this field, which is in the same format as the original NRCS Linesource field. Populate the field by using one or more of the standardized codes listed below in uppercase. If more than one code is used, then separate the values with a comma and no space with the most recent LineSource used listed first in the sequence. Other reference and source maps not listed should be noted in the metadata (section 6.2.4).

TOPO24   Delineated from hardcopy 1:24,000-scale topographic maps.

TOPO25   Delineated from hardcopy 1:25,000-scale topographic maps for the Caribbean outlying areas.

TOPO50   Delineated from hardcopy 1:25,000-scale topographic maps.

TOPO63   Delineated from hardcopy 1:63,360-scale topographic maps for Alaska.

TOPO250   Delineated from hardcopy 1:250,000-scale topographic maps.

DRG24   Delineated from 1:24,000-scale Digital Raster Graphics.

DRG25   Delineated from 1:25,000-scale Digital Raster Graphics for the Caribbean outlying areas.

DRG63   Delineated from 1:63,360-scale Digital Raster Graphics for Alaska.

CAN50   Delineated from 1:50,000-scale Digital Raster Graphics (CanMatrix).

CAN250   Delineated from 1:250,000- scale Digital Raster Graphics (CanMatrix).

CDED50   Delineated from 1:50,000- scale Canadian Digital Elevation Data.

CDED250   Delineated from 1:250,000- scale Canadian Digital Elevation Data.

DEM10   Derived from 10-meter Digital Elevation Model.

DEM15   Derived from 15-meter Digital Elevation Model.

DEM30   Derived from 30-meter Digital Elevation Model.

NED10   Derived from 10-meter National Elevation Dataset Model.

NED30   Derived from 30-meter National Elevation Dataset Model.

NOAA3NM   National Oceanic and Atmospheric Administration demarcation 3 nautical miles offshore. May be generalized by the user in complex configuration, such as islands where buffer overlaps.

EDNA30   Derived from 30-meter Elevation Derivatives for National Applications.

BATH "scale"   Interpreted from NOAA 1:24,000-scale or other bathymetric data, for example, BATH24.

BUFFER "distance" "unit"   A near offshore limit used in bays, sounds or estuaries. The distance offshore is determined by local groups in the State. For unit, use: "M" for meter, "F" for feet, or "N" for nautical mile; for example, BUFFER400F.

NAIP "year"   Delineated from aerial photography produced by the National Agriculture Imagery Program, for example, NAIP2005

NHD24   Interpreted from 1:24,000-scale National Hydrography Dataset.

NHD25   Interpreted from 1:25,000-scale National Hydrography Dataset.

NHD63   Interpreted from 1:63,360-scale National Hydrography Dataset.

NHD100   Interpreted from 1:100,000-scale National Hydrography Dataset.

NHN10   Interpreted from 1:10,000-scale National Hydrography Network (Canada).

NHN20   Interpreted from 1:20,000-scale National Hydrography Network (Canada).

NHN50   Interpreted from 1:50,000-scale National Hydrography Network (Canada).

HYPSO "scale"   Delineated from 1:24,000-scale or other contour data, for example, HYPSO24.

ORTHO "scale"   Interpreted from 1:12,000-scale or other orthoimagery, for example, ORTHO12.

DEDEM10   Drainage-enforced 10-meter Digital Elevation Model.

DEDEM30   Drainage-enforced 30-meter Digital Elevation Model.

GPS   Derived from Global Positioning System.

NRN   Interpreted from GPS transportation data captured at 10-meter accuracy (Canada).

LIDAR   Derived from LiDAR (light detection and ranging) data. See required metadata form for details on resolution and acquisition.

IFSAR   Derived from IfSAR (interferometric synthetic aperture radar) data.

TRIM   Delineated from 1:20,000-scale British Columbia Terrain Resource Information Management three-dimensional data.

"Province" "scale"   Derived from Provincial GIS layers. Specify with a 2-digit Province abbreviation followed by scale, for example use MB20 for Manitoba 1:20,000-scale data.

SPOT "year"   Interpreted from System for Earth Observation imagery.

CAN   Various Canadian base reference information was used. See FGDC metadata for more information.

MEX   Various Mexican base reference information was used. See FGDC metadata for more information.

OTH   Other.

If additional information is required, it can be stored in the accompanying FGDC metadata.

Field Name	LineSource
Field Type	Character
Field Width	30
Domain	TOPO24, TOPO25, TOPO50, TOPO63, TOPO250, DRG24, DRG25, DRG63, CAN50, CAN250, CDED50, CDED250, DEM10, DEM15, DEM30, NED10, NED30, NOAA3NM, EDNA30, BATH "scale," NOAA3NM, BUFFER "distance" "unit," NAIP "year," NHD24, NHD25, NHD63, NHD100, NHN10, NHN20, NHN50, HYPSO "scale," ORTHO "scale," DEDEM10, DEDEM30, GPS, NRN, LIDAR, IFSAR, TRIM, "Province" "scale," SPOT "year," CAN, MEX, OTH
Value required?	Yes
Example	DRG24

6.2.2.6.5   WBDLine: LoadDate

LoadDate automatically generates and assigns the date when the data were loaded into the official USGS WBD ArcSDE database. This field is the effective date for all feature edits.

Field Name	LoadDate
Field Type	Date
Field Width	22
Domain	Calculated system date
Value required?	Yes
Example	9/21/2011 10:41:40 AM

6.2.2.7   WBDPoint

WBDPoint represents the centroid of each Polygon Unit and is the link between the WBDAttributes table and the derived polygons. Because some multipart hydrologic units in the WBD share the same attribution information, WBDPoint includes five common fields and four unique fields described below.

6.2.2.7.1   Common Fields

Like WBDLine, WBDPoint contains PermanentIdentifier, SourceFeatureID, MetaSourceID, SourceDataDesc, SourceOriginator fields. These common fields are automatically created and assigned. See the previous descriptions of these fields (sections 6.2.2.1–6.2.2.5).

6.2.2.7.2   WBDPoint: HUC

HUC is a text field, completed by the WBD In-State Steward, which contains the 2-, 4-, 6-, 8-, 10, 12-, 14-, or 16-digit hydrologic unit code. This code is unique to each hydrologic unit at each nested level. The HUC codes use the number of digits to represent the level of the hydrologic unit. Each subsequent level gains two additional digits as it moves through the hierarchy.

Field Name	HUC
Field Type	Character
Field Width	16
Domain	(2-, 4-, 6-, 8-, 10-, 12-, 14-, or 16-digit hydrologic unit code, nested by area)
Value required?	Yes
Example	1019000401

6.2.2.7.3   WBDPoint: HULevel

HULevel is a domain-based field, completed by the WBD In-State Steward, which expresses the WBD line level as a single digit. For example, a 12-digit hydrologic unit line would be a level 6 line, and a 4-digit hydrologic unit line would be a level 2 line. This field allows for crosswalking linked reports and external data that utilized the level classification.

Field Name	HULevel
Field Type	Long Integer
Field Width	6
Domain	1, 2, 3, 4, 5, 6, 7, 8
Value required?	Yes
Example	6

6.2.2.7.4   WBDPoint: HUClass

HUClass is a domain-based field, completed by the WBD In-State Steward, which indicates the minimum number of digits of the hydrologic unit that the line bounds.

Field Name	HUClass
Field Type	Long Integer
Field Width	16
Domain	2, 4, 6, 8, 10, 12, 14, 16
Value required?	Yes
Example	12 (subwatershed)

6.2.2.7.5   WBDPoint: LoadDate

LoadDate represents the date when the data were loaded into the official USGS WBD ArcSDE database. The field is the effective date for all feature edits, and it is automatically generated.

Field Name	LoadDate
Field Type	Date
Field Width	22
Domain	(Automatically calculated system date)
Value required?	Yes
Example	9/21/2011 10:41:40 AM

6.2.2.8   WBDAttributes

WBDAttributes is the table where the hydrologic unit code, gazetteer identification, hydrologic unit name, hydrologic unit class, hydrologic unit type, noncontributing area, in acres, and load date are stored. This table also contains a permanent feature identification, provides a link to the metadata tables, and is used for the attributes of the derived polygon classes. WBDAttributes includes five common fields and seven unique fields described below.

6.2.2.8.1   Common Fields

Like WBDLine and WBDPoint, WBDAttributes contains PermanentIdentifier, SourceFeatureID, MetaSourceID, SourceDataDesc, and SourceOriginator fields. These fields are automatically generated. See the previous descriptions of these fields (sections 6.2.2.1–6.2.2.5).

6.2.2.8.2   WBDAttributes: HUC

HUC is a text field, completed by the WBD In-State Steward, which contains the 2-, 4-, 6-, 8-, 10-, 12-, 14-, or 16-digit hydrologic unit code. This code is unique to each hydrologic unit at each nested level. Each subsequent level gains two additional digits as it moves through the hierarchy.

Field Name	HUC
Field Type	Character
Field Width	16
Domain	(2-, 4-, 6-, 8-, 10-, 12-, 14-, or 16-digit hydrologic unit code, nested by area)
Value required?	Yes
Example	1019000401

6.2.2.8.3   WBDAttributes: GazID

GazID is an automatically generated number field that uses a unique number to relate the name of the hydrologic unit to the GNIS names database. This field is not populated by the WBD In-State Steward.

Field Name	GazID
Field Type	Long Integer
Field Width	16
Domain	(Unique sequential number assigned by GNIS)
Value required?	Yes
Example	123456789

6.2.2.8.4   WBDAttributes: Name

The name field contains the assigned name of the hydrologic unit, constructed with reference to a feature name that is officially recognized by the GNIS. The WBD In-State Steward completes this field, based on the rules described in section 5.3, "Naming Protocol for 10-, 12-, 14-, and 16-Digit Hydrologic Units." The value assigned to the 10- and 12-digit name attribute should be unique within the 8-digit hydrologic units that contain the10- and 12-digit hydrologic unit.

Field Name	Name
Field Type	Character
Field Width	255
Domain	(GNIS official names)
Value required?	Yes
Example	Gerome Creek

6.2.2.8.5   WBDAttributes: HUClass

HUClass is a domain-based field, completed by the WBD In-State Steward, which indicates the number of digits in the hydrologic unit code assigned to the hydrologic unit.

Field Name	HUClass
Field Type	Long Integer
Field Width	16
Domain	2, 4, 6, 8, 10, 12, 14, 16
Value required?	Yes
Example	12 (subwatershed)

6.2.2.8.6   WBDAttributes: HUType

HUType contains a letter code that defines the type of hydrologic unit. This code is used only in the 10-, 12-, 14-, or 16-digit hydrologic units. The HUType attribute is the single-letter abbreviation from the list of official names provided below. The HUType field is completed by the WBD In-State Steward. Use the single type that most closely describes the 12-digit hydrologic unit (see coastal examples in fig. 10).

Field Name	HUType
Field Type	Text
Field Width	2
Domain	S, C, F, M, W, I
Value required?	No
Example	S (standard)

S "Standard" hydrologic unit   An area with drainage flowing to a single outlet point, excluding noncontributing areas. Some examples include "true," "standard," "composite," and "remnant" hydrologic units (section 3.5.1 and section 3.5.2).

C "Closed Basin" hydrologic unit   A drainage area where all surface flow is internal (100-percent noncontributing); no overland flow leaves the hydrologic unit through the outlet point (section 3.5.3).

F "Frontal" hydrologic unit   An area along the coastline of a lake, ocean, bay, etc. that has more than one outlet. These hydrologic units are predominantly land; however, they may include some water areas at or near the outlet(s) (section 3.6.2).

M "Multiple Outlet" hydrologic unit   An area that has more than one natural outlet, such as an outlet located on a stream that has multiple channels (for example, braided streams, deltas or alluvial fans). Multiple Outlet does not include frontal or water hydrologic units, hydrologic units with artificial inter-basin transfers, drainage outlets through karst or groundwater flow, or outlets that cross a stream with an island. Use this code only in rare instances.

W "Water" hydrologic unit   An area that is predominantly water with adjacent land areas; for example, a lake, estuary, or harbor (section 3.6).

I "Island" hydrologic unit   An area that is one or more islands, which includes surrounding water (section 3.6.3).

6.2.2.8.7   WBDAttributes: NContrbAcres

NContrbAcres is the area, in acres, of a hydrologic unit that does not drain outside of its own boundaries, in other words, the noncontributing area attribute represents the area that does not contribute to downstream accumulation of streamflow under normal flow conditions (section 3.5.3). If a noncontributing area is on the boundary between two or more hydrologic units, then determine the low point along the non-contributing area boundary, and associate the noncontributing area with the hydrologic unit adjacent to the low point on the boundary. The value is the total acreage of the noncontributing areas within a hydrologic unit. The WBD In-State Steward completes this field.

Field Name	NContrbAcres
Field Type	Double
Field Width	16
Domain	0–unlimited
Value required?	No
Example	367

6.2.2.8.8   WBDAttributes: LoadDate

LoadDate represents the date when the data were loaded into the official USGS WBD ArcSDE database. This automatically generated field is the effective date for all feature edits.

Field Name	LoadDate
Field Type	Date
Field Width	22
Domain	(Automatically calculated system date)
Value required?	Yes
Example	9/21/2011 10:41:40 AM

6.2.2.9   WBDPointEvent

This feature resides and is managed in the NHD.

6.2.2.10   WBDNav

WBDNav is the flow table for the WBD, providing the downstream link for each hydrologic unit from its upstream unit or units. This table is also used to show main flow up- and downstream where multiple units flow into a single unit or where a single unit flows into multiple units.

6.2.2.10.1   WBDNav: FromHUC

FromHUC is an automatically generated 2-, 4-, 6-, 8-, 10-, 12-, 14-, or 16-digit code that represents the originating hydrologic unit.

Field Name	FromHUC
Field Type	Text
Field Width	16
Domain	(Automatically calculated)
Value required?	Yes
Example	101900040101

6.2.2.10.2   WBDNav: ToHUC

ToHUC is a 2- to 16-digit number representing the hydrologic unit into which the FromHUC unit flows. The WBD In-State Steward completes this field only for the 12-digit hydrologic unit.

Field Name	ToHUC
Field Type	Text
Field Width	16
Domain	None
Value required?	No
Example	101900040102

6.2.2.10.3   WBDNav: HUClass

HUClass is a domain-based field, completed by the WBD In-State Steward, which indicates the number of digits that the FromHUC (see section 6.2.2.10.1) hydrologic unit contains.

Field Name	HUClass
Field Type	Long Integer
Field Width	16
Domain	(2-, 4-, 6-, 8-, 10-, 12-, 14-, 16-digit valid hydrologic unit codes)
Value required?	Yes
Example	12 (subwatershed)

6.2.2.10.4   WBDNav: UpMain

UpMain is the main upstream hydrologic unit from the value listed in FromHUC (see section 6.2.2.10.1). The WBD In-State Steward completes this field. This capability is planned and not currently available.

Field Name	UpMain
Field Type	Long Integer
Field Width	16
Domain	Yes, No
Value required?	No
Example	Yes

6.2.2.10.5 WBDNav: DownMain

DownMain is the downstream hydrologic unit from the value listed in FromHUC (see section 6.2.2.10.1). The WBD In-State Steward completes this field. This capability is planned and not currently available.

Field Name	DownMain
Field Type	Long Integer
Field Width	16
Domain	Yes, No
Value required?	No
Example	Yes

6.2.2.11 WBDFeatureToHUMod

WBDFeatureTo HUMod is a table that stores in a domain all of the modification attributes of lines and polygons, so that database integrity is maintained. The line and polygon modifications, described below, are stored in this table and related back to their respective WBDLine and hydrologic units by the PermanentIdentifier.

6.2.2.11.1 WBDFeatureToHUMod for lines: HUMod

The HUMod line modification attribute, completed by the WBD In-State Steward, is the two-character, uppercase abbreviation(s) code used to track either the modification to natural overland flow that alters the location of the hydrologic unit boundary or special conditions that are applied to a specific boundary line segment. The value identifies the type of modification, from the list provided, that has been applied to the boundary segment. If more than one abbreviation is used, then separate them by commas without spaces and list them in order of importance.

Field Name	HUMod
Field Type	Character
Field Width	20
Domain	AW, DM, LA, LE, MA, NM, OC, OF, PD, PL, PS, SI, SL, TF, UA
Value required?	Yes
Example	PD, PS

AW Artificial Waterway   An aqueduct, canal, ditch or drain used to transport surface water, altering the natural flow out of the hydrologic unit. (Options previously included AD Aqueduct, DD Drainage Ditch, GC General Canal/Ditch, ID Irrigation Ditch, IT Interbasin Transfer, SD Stormwater Ditch, SC Stormwater Canal, and BC Barge Canal, which are now grouped into this designation).

DM Dam   A barrier constructed to control the flow or raise the level of water at a hydrologic unit outlet or on the hydrologic unit boundary; alters the natural boundary location.

LA Lava Field   A lava field is a large expanse of nearly flat lying lava flows (usually deposited in the past).

LE Levee   An artificial bank to confine a stream channel or limit adjacent areas subject to flooding; alters the natural boundary location.

MA Mining Activity   Heavy topographic modification of a hydrologic unit by surface mining; alters natural boundary location.

NM No Modifications   No modifications are present. Use when no other options with the modification domain have been cited.

OC Overflow Channel or Flume   An artificial channel built to control excess high flow from a natural channel; alters the natural boundary location.

OF Overbank Flow   A natural condition in which a stream surpasses bankfull stage and the excess flows into a nearby channel draining to a different hydrologic unit (special condition; see example in fig. 18).

PD Pipe Diversion   A redirection of surface water by a pipeline from one hydrologic unit to another; alters the natural boundary location.

PL Playa   A playa is sandy, salty or mud-caked flat floor of a dessert basin having interior drainage usually occupied by a shallow lake during or after prolonged, heavy rains.

PS Pumping Station   A facility along a stream or other waterbody used to move water over a levee or other obstruction; alters the natural boundary location.

SI Siphon   An artificial diversion, which is usually named "Siphon" on maps, to move surface water from one stream channel to another; alters the natural boundary location.

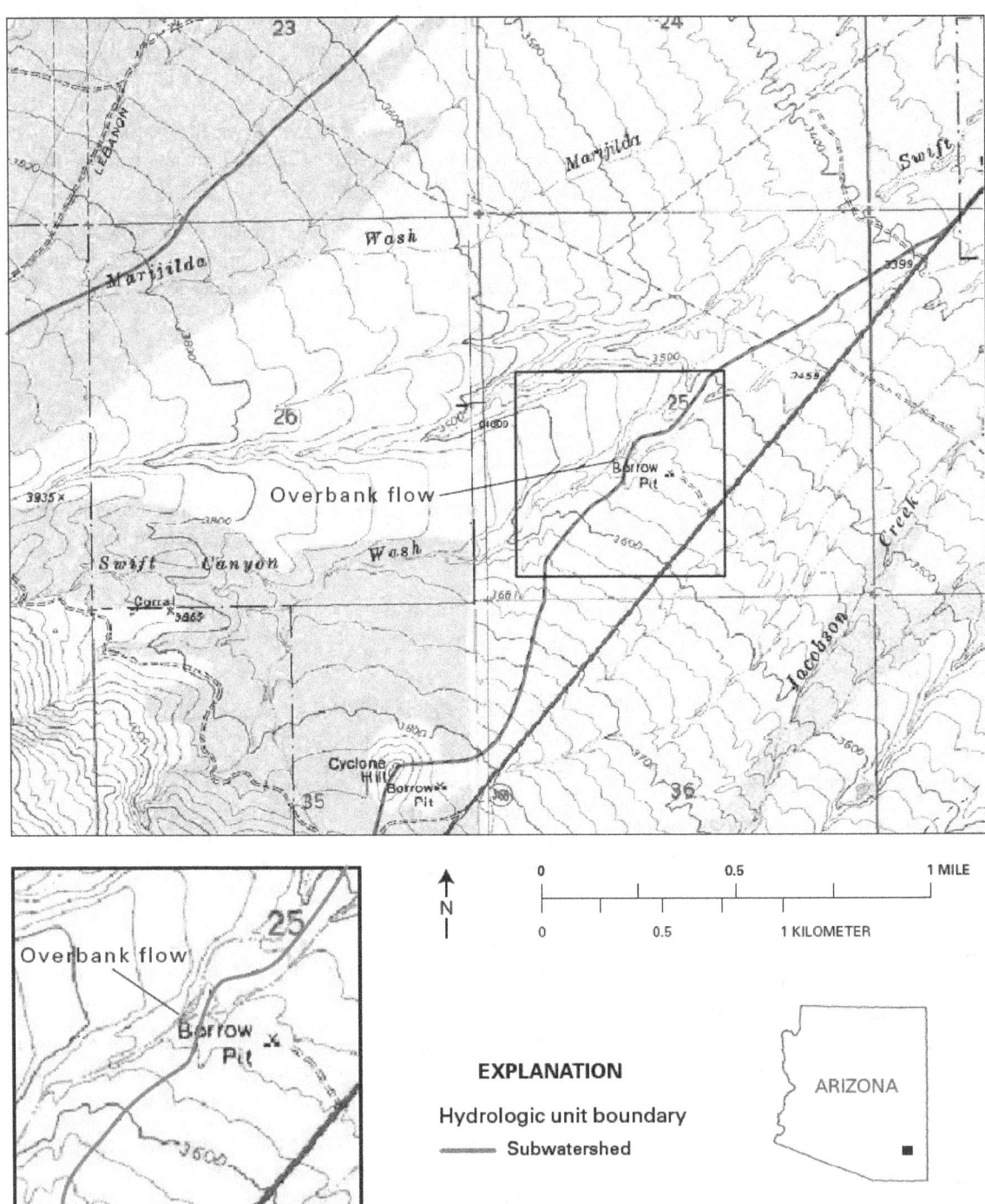

Figure 18.   When high flows cause a natural diversion and the excess runs into an adjacent channel draining to a different hydrologic unit, the special condition is named overbank flow. Overbank flow is usually found in areas of alluvial material, such as in this area of Graham County, Arizona.

SL Shoreline   A demarcation of shoreline within WBD agreed upon by in-State groups. These instances are exceptions to the preferred method.

TF Transportation Feature   A land transportation feature, for example, a road, railroad, dock, airport, etc., that alters the natural boundary location.

UA Urban Area   Heavy modification of hydrologic unit topography by development beyond that described above in "Transportation Feature" that alters the natural boundary location.

6.2.2.11.2   WBDFeatureTo HUMod for polygons: HUMod

The HUMod 12-digit hydrologic unit polygon modification attribute, entered by the WBD In-State Steward, is a two-character, uppercase abbreviation for either (1) the type of modification to natural overland flow that alters the natural delineation of a 12-digit hydrologic unit or (2) the special conditions GF-groundwater flow, GL-glacier, IF-ice field, KA-karst, and NC-noncontributing area. The value of the HUMod field helps to indicate the areal location of the modification to the 12-digit hydrologic unit. If more than one abbreviation is used, then separate them by commas and list them in order of importance.

Previous versions of this guideline did not provide the same number of modification choices—check metadata or contact the WBD In-State Steward for more information.

Field Name	HUMod
Field Type	Character
Field Width	20
Domain	AW, GF, GL, IF, KA, LA, MA, NC, NM, OC, OF, PD, RC, RS, UA, WD
Value required?	Yes
Example	NC

AW Artificial Waterway   A canal, ditch, or drain used to transport surface water that alters the natural flow out of the hydrologic unit. (The previously included designations AD Aqueduct, DD Drainage Ditch, GC General Canal/Ditch, ID Irrigation Ditch, IT Interbasin Transfer, SD Stormwater Ditch, SC Stormwater Canal, and BC Barge Canal have now been grouped into this designation). Withdrawing and receiving hydrologic units should carry this designation, as well as all hydrologic units in which the flow is altered by an artificial waterway.

GF Groundwater Flow   A special condition, usually in locations with sandy soil, where most of the runoff in a hydrologic unit drains underground.

GL Glacier   A special condition where a hydrologic unit crosses or includes a body or stream of ice moving outward and downslope from an area of accumulation; area of relatively permanent snow or ice on the top or side of a mountain or mountainous area.

IF Ice Field   A special condition where a hydrologic unit crosses or includes a field of ice, formed in regions of perennial frost.

KA Karst   A special condition where a hydrologic unit is within an area of, or includes an area of, geologic formations of irregular limestone deposits with sinks, underground streams, or caverns.

LA Lava Field   A lava field is a large expanse of nearly flat lying lava flows (usually deposited in the past).

MA Mining Activity   Heavy topographic modification of a hydrologic unit by surface mining; alters the natural flow out of the hydrologic unit.

NC Noncontributing Area   A naturally formed area that does not contribute surface-water runoff to a hydrologic unit outlet under normal conditions, for example, a playa. This does not include groundwater flow.

NM No Modifications   No modifications are present. Use when no other options with the modification domain have been cited.

OC Overflow Channel or Flume   An artificial channel built to control excess high flow from a natural channel; alters the natural flow out of the hydrologic unit.

OF Overbank Flow   A natural condition in which a stream surpasses bankfull stage and the excess flows into a nearby channel draining to a different hydrologic unit (special condition; see example in fig. 18). Both the losing and the receiving hydrologic unit should carry this designation.

PD Pipe Diversion   A redirection of surface water by a pipeline from one hydrologic unit to another; alters the natural flow out of the hydrologic unit.

RC Receiving   A hydrologic unit that receives diverted water.

RS Reservoir   A constructed basin formed to contain and store water for future use in an artificial lake; alters the natural flow out of the hydrologic unit.

UA Urban Area   Heavy modification of hydrologic unit topography by development beyond that described in section 6.2.11.1, "Transportation Feature," that alters natural flow out of the hydrologic unit.

WD Withdrawal   A hydrologic unit from which water is diverted.

6.2.2.12   NWISLine

NWISLine is a feature dataset used for storing lines needed for the NWIS drainage areas but that are not part of the WBD.

6.2.3   Derived Classes

The WBD derived feature classes are probably the most familiar and frequently used WBD feature classes. These layers are "derived" because they are not edited by the WBD In-State Stewards; rather, they are derived from the edited data. The data are derived from the point and line features as well as the attribute and modification tables. Each time edits have been reviewed by the WBD–NTC and are successfully incorporated into the WBD, those areas where the edits took place are rederived and posted for downloading.

The following attributes are derived by the system:

AreaAcres   This value is calculated from the intrinsic area value maintained by the GIS software; therefore, acreage values may vary from user calculations, depending on the projection of the data. North America Albers Equal Area Conic, North American Datum 1983 is the recommended projection. If the units of the area field are stored in square meters, then use the conversion factor 0.0002471. For example, 40,469,446 square meters multiplied by 0.0002471 = 10,000 acres.

AreaSqKm   This value is calculated from the intrinsic area value maintained by the GIS software; therefore, the square kilometers values may vary from user calculations, depending on the projection of the data. North America Albers Equal Area Conic, North American Datum 1983 is the recommended projection.

States   The State or outlying area attribute identifies State(s) or outlying areas that the hydrologic unit falls within or touches. The U.S. Census Bureau 1:100,000-scale State layer will be used to establish State boundaries in the derivation process. The two-letter U.S. Postal Service State abbreviation is assigned. If a hydrologic unit crosses into Canada, the two-letter Canada Post abbreviation for the Province is used. If a unit crosses into Mexico, the MX designation is used. If more than one abbreviation is used, they are separated by commas, without spaces, and are sorted in alphabetical order.

LoadDate   A system-generated date that is assigned upon posting accepted edits to the national official database.

6.2.3.1   WBDHU2

WBDHU2 represents the 2-digit hydrologic unit boundaries (previously referred to as Regions). There are 22 2-digit hydrologic units (Regions) in the WBD, and each has the following attribute fields.

- **GazID**   Gazetteer identification number (also referred to as GNIS)
- **AreaAcres**   Area, shown in acres
- **AreaSqKm**   Area, shown in square kilometers
- **States**   States included in the 2-digit hydrologic unit
- **LoadDate**   Date when the data were loaded
- **HU2Name**   Region name
- **HUC2**   Unique 2-digit hydrologic unit code

6.2.3.2   WBDHU4

WBDHU4 represents the 4-digit hydrologic unit boundaries (previously referred to as Subregions). There are 220 4-digit hydrologic units (Subregions) in the WBD, and each has the following attribute fields.

- **GazID**   Gazetteer identification number (also referred to as GNIS)
- **AreaAcres**   Area, shown in acres
- **AreaSqKm**   Area, shown in square kilometers
- **States**   States included in the 4-digit hydrologic unit
- **LoadDate**   Date when the data were loaded
- **HU4Name**   Subregion name
- **HUC4**   Unique 4-digit hydrologic unit code

6.2.3.3 WBDHU6

WBDHU6 represents the 6-digit hydrologic unit boundaries (previously referred to as Basins). There are approximately 378 6-digit hydrologic units (Basins) in the WBD, and each has the following attribute fields.

- **GazID** Gazetteer identification number (also referred to as GNIS)

- **AreaAcres** Area. shown in acres

- **AreaSqKm** Area. shown in square kilometers

- **States** States included in the 6-digit hydrologic unit

- **LoadDate** Date when the data were loaded

- **HU6Name** Basin name

- **HUC6** Unique 6-digit hydrologic unit code

6.2.3.4 WBDHU8

WBDHU8 represents the 8-digit hydrologic unit boundaries (previously referred to as Subbasins). There are approximately 2,283 8-digit hydrologic units (Subbasins) in the WBD, and each has the following attribute fields.

- **GazID** Gazetteer identification number (also referred to as GNIS)

- **AreaAcres** Area, shown in acres

- **AreaSqKm** Area, shown in square kilometers

- **States** States included in the 8-digit hydrologic unit

- **LoadDate** Date when the data were loaded

- **HU8Name** Subbasin name

- **HUC8** Unique 8-digit hydrologic unit code

6.2.3.5 WBDHU10

WBDHU10 represents the 10-digit hydrologic unit boundaries (previously referred to as Watersheds). There are approximately 17,828 10-digit hydrologic units (Watersheds) in the WBD, and each has the following attribute fields.

- **GazID** Gazetteer identification number (also referred to as GNIS)

- **AreaAcres** Area, shown in acres

- **AreaSqKm** Area, shown in square kilometers

- **States** States included in the 10-digit hydrologic unit

- **LoadDate** Date when the data were loaded

- **HU10Name** Watershed name

- **HU10Type** Watershed type

- **HUC10** Unique 10-digit hydrologic unit code

6.2.3.6 WBDHU12

WBDHU12 represents the 12-digit hydrologic unit boundaries (previously referred to as Subwatersheds). There are approximately 97,443 12-digit hydrologic units (Subwatersheds) in the WBD, and each has the following attribute fields.

- **GazID** Gazetteer identification number (also referred to as GNIS)

- **AreaAcres** Area, shown in acres

- **AreaSqKm** Area, shown in square kilometers

- **States** States included in the 12-digit hydrologic unit

- **LoadDate** Date when the data were loaded

- **HU12Name** Subwatershed name

- **HU12Type** Subwatershed type

- **HUC12** Unique 12-digit hydrologic unit code

- **HU12Mod** Polygon modifications

- **NContrbAcres** Noncontributing area, in acres

- **NcontrbSqKm** Noncontributing area, in square kilometers

6.2.3.7   WBDHU14

WBDHU14 represents the 14-digit hydrologic unit boundaries. There is no common name for this hydrologic unit. This is a new subdivision in the WBD and data have yet to be added to it. Each 14-digit hydrologic unit has the following attribute fields.

- **GazID**   Gazetteer identification number (also referred to as GNIS)

- **AreaAcres**   Area, shown in acres

- **AreaSqKm**   Area, shown in square kilometers

- **States**   States included the 14-digit hydrologic unit

- **LoadDate**   Date when the data were loaded

- **HU14Name**   14-digit name

- **HU14Type**   14-digit type

- **HUC14**   Unique 14-digit hydrologic unit code

- **HU14Mod**   Polygon modifications

- **NContrbAcres**   Noncontributing area, in acres

- **NcontrbSqKm**   Noncontributing area, in square kilometers

6.2.3.8   WBDHU16

WBDHU16 represents the 16-digit hydrologic unit boundaries. There is no common name for this hydrologic unit. This is a new subdivision in the WBD, and data have yet to be added to it. Each 16-digit hydrologic unit has the following attribute fields.

- **GazID**   Gazetteer identification number (also referred to as GNIS)

- **AreaAcres**   Area, shown in acres

- **AreaSqKm**   Area, shown in square kilometers

- **States**   States included in the 16-digit hydrologic unit

- **LoadDate**   Date when the data were loaded

- **HU16Name**   16-digit name

- **HU16Type**   16-digit type

- **HUC16**   Unique 16-digit hydrologic unit code

- **HU16Mod**   Polygon modifications

- **NContrbAcres**   Noncontributing area, in acres

- **NcontrbSqKm**   Noncontributing area, in square kilometers

6.2.3.9   WBDNWIS

WBDNWIS is the newest feature class in the WBD, and it represents the drainage area boundaries of gaging stations stored within the NWIS database.

6.2.4   Metadata

Metadata in the new model are kept not only at the dataset level but also at the feature level. Each time an item is updated or edited in any way, metadata are required to show how, when, and why the feature was modified. The WBD Edit Tool will not allow editing without correctly populated metadata. The WBD Edit Tool provides a template and tool needed to create metadata properly.

6.2.4.1   MetaProcessDetail

MetaProcessDetail is the template that defines the overarching process related to the associated set of edits and to provide contact information for and instructions about questions of those specific edits (fig. 19).

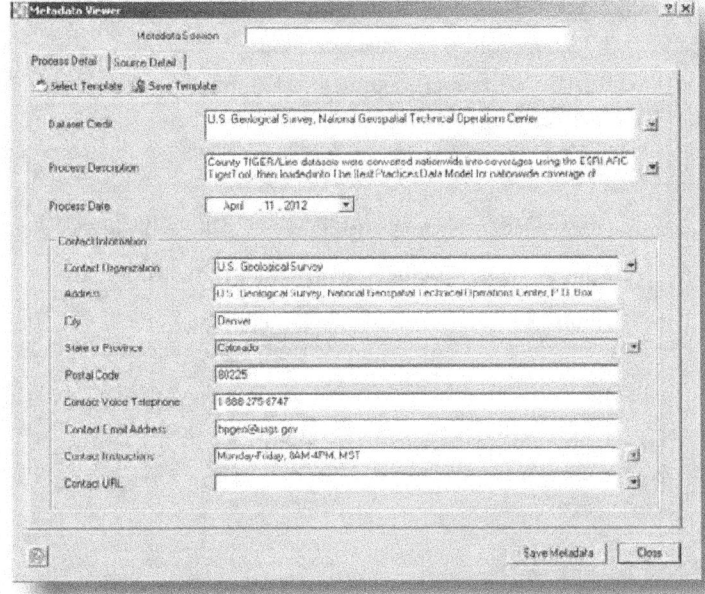

Figure 19.   The Metadata Process Details template within the WBD Edit Tool is where information is entered about the editor of a particular WBD edit and why the edit was made. This example shows the contact information portion of the template, should an end user need more information about a particular change.

6.2.4.2   MetaSourceDetail

MetaSourceDetail is the metadata template where information is entered about the base data used to make the edits and updates to the WBD (fig. 20). This information includes the type of data (LiDAR, DEM, DRG, NAIP), the source of the data, and the creator of the data, the relevant dates associated with the creation of the data, the scale and accuracy of the base data, and a brief description of the source data.

6.2.4.3   MetaID

Prior to integration of the WBD with the NHD, the metadata identification attribute was the code used for tracking changes made to a specific boundary (line or arc) segment or polygon attribute. When a State received provisional certification status, all boundary segments and polygon attribute values were 01. After provisional certification, the MetaID value was incrementally updated for that feature whenever changes were made to a boundary segment or polygon attribute. The change was documented in the updated FGDC metadata, and it referred back to the MetaID.

The four-character MetaID began with the two-letter U.S. Postal Service State abbreviation, in uppercase, followed by a two-digit sequence number; for example, "OK01," "ID02." All State polygons had a MetaID of (State abbreviation 01) at the time of provisional certification. Appendix 11.5 documents the previously used method for tracking metadata.

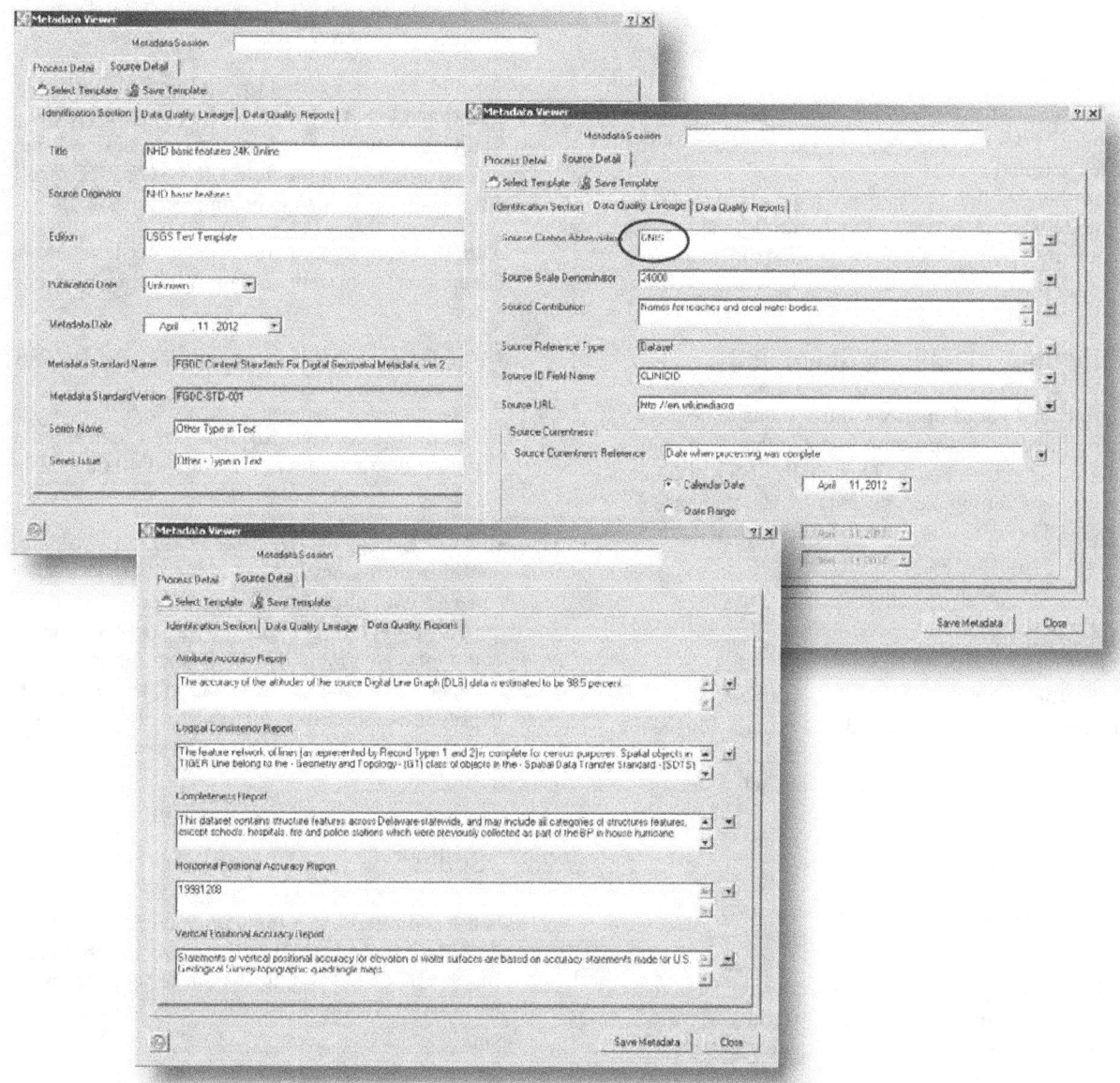

Figure 20.   The Metadata Source Detail template within the WBD Edit Tool is where information is entered about the base data used to make the edits and updates to the WBD. The example shows that GNIS data is being used to help name a hydrologic unit.

7. Quality Assurance and Quality Control

7.1 Quality Assurance

Hydrologic units are delineated and verified through an interagency process. Before newly delineated hydrologic unit boundaries are submitted for the WBD, the boundaries must be reviewed for conformance to these guidelines by the WBD In-State Steward originating agency and/or designated members of an interagency hydrologic unit coordinating group within the State. Qualified reviewers typically would be hydrologists or natural resource and GIS specialists with background and experience in hydrologic unit delineation. Representatives of the interagency hydrologic unit coordinating group and/or regional/local parties should participate in the development and review of delineations before the data are submitted for national review for compliance with the required standard and release to the public. The WBD In-State Steward should thoroughly review the data before submitting it to the WBD NTC for national review.

It is recommended that reviews and edit checks be done throughout the delineation process. At a minimum, edit checks should be made after the hydrologic units are delineated, mapped, and digitized.

7.2 Quality Control (Editing Checklist)

A recommended list of quality-control items to be checked during the delineation, coding, documentation, and digitizing process is shown below. This quality-control process is necessary for maintaining a consistent national database of hydrologic units. This is not an exhaustive list, but it covers most of the items requiring verification.

7.2.1 Delineation

- Do the hydrologic unit boundaries match at State boundaries, and are coding and size criteria consistent across boundaries? Resolve differences in delineation at State boundaries before the data are submitted for agency certification.

- Does linework meet current NSSDA?

- Are hydrologic units consistently delineated across the State?

- Are the hydrologic units correctly delineated against the minimum required scale NRCS County Mosaic DRGs with respect to hydrography and elevation contours? (See section 4.)

- Is the NHD shoreline representation the basis for distance-based nearshore delineation? If not, explain what was used and why.

- Do coastal delineations in bays, sounds, and estuaries meet the guidance criteria?

  - Are open-ocean boundaries complete where bays, sounds, and estuaries do not predominate?

  - Are 8-digit hydrologic units extended to the open-ocean NOAA Three Nautical Mile Line?

- Do hydrologic units that were difficult to delineate include a description of the procedure used? Make notes about how certain boundaries were established or field checked so that they can be documented in the field LINESOURCE (section 6.2.2.6.4) and in the metadata (section 6.2.4).

- Were local sources used to resolve the locations of questionable boundaries? Make notes where local knowledge supersedes source maps, and document this information in the field LINESOURCE (section 6.2.2.6.4) and in the metadata (section 6.2.4).

- Although number and area criteria per level are listed, exceptions for extreme geomorphology will be considered.

  - Are terrestrial hydrologic units subdivided into 5 to 15 units? (See section 3.3.)

  - Are the areas of terrestrial 10-digit hydrologic units within the recommended range of 40,000–250,000 acres?

  - Are the areas of terrestrial 12-digit hydrologic units within the recommended range of 10,000–40,000 acres, with none less than 3,000 acres? When a hydrologic unit is divided by a State boundary, check the total area including both States' portion. (See section 3.4.) A combined total of 10 percent of the polygons (combined 10- and 12-digit hydrologic units) outside the specified size criteria is allowed for the State.

7.2.2 Codes

- Is the coding of hydrologic units consistent within an 8-digit hydrologic unit that covers more than one State? Check the coding system for correctness, duplication, or missing codes. Each polygon must have a single label. (See section 5.2.)

- Does the coding of hydrologic units within a given level meet the guidelines for starting upstream and progressing downstream? (See section 5.2.)

7.2.3 Attributes

- Are the attribute fields complete, with valid values? (See section 6.2.2.8)

- Do polygons have the required values in these fields?

  - HUClass, HUC (at both the 10- and 12-digit, and 14- and 16-digit, if applicable; see section 6.2.2.8.5)

  - Name (at both the 10- and 12-digit, and 14- and 16-digit, if applicable; see section 6.2.2.8.4)

  - HUMod, HUType (at both the 10- and 12-digit, and 14- and 16-digit, if applicable; see sections 6.2.2.8.6 and 6.2.2.11.1)

- Do lines have the required values in these fields?

  - HUClass (see section 6.2.2.8.5)

  - HULevel (see section 6.2.2.7.3)

  - Linesource (see section 6.2.2.6.4)

  - HUMod (See section 6.2.2.11.1)

- Do coastal hydrologic units have the correct values in these fields?

  - HULevel (see section 6.2.2.6.2)

- Do artificial modifications or special conditions exist? Check and document in the fields HUMod for both lines and polygons, if the natural boundary location has been modified or special conditions exist. (See section 6.2.2.11.1.)

7.2.4 Data Format

Does the file format conform to the guidelines? Before submitting data for national review and certification, verify that the coordinate system used in the data is correct. Data submitted for certification must be in geographic coordinates stored in decimal degrees. The horizontal datum is the North American Datum of 1983 (NAD 83). The data should be in double-precision accuracy.

8. Data Revision, Checkout, Editing, Verification, and Submittal

This section presents guidelines and procedures to be followed for successful checkout, editing, and submittal of State contributions to the WBD. Digital geospatial data and attributes must match adjoining States. State submissions to the WBD will be made by use of the WBD stewardship Web site, using provided tools.

8.1 State Data Revision

Agencies involved in the development of the WBD should establish organizational responsibility for reviewing the data before submittal. After the revisions are completed, the WBD In-State Steward will submit the data to USGS through the WBD Edit Tools for review and assurance of compliance with the standards by the WBD–NTC. Supporting material, such as text documents describing the procedure used to delineate and digitize the hydrologic units, is documented in the LINESOURCE field and metadata. In addition, a summary of decisions for placement of hydrologic unit boundaries in complex hydrographic areas and access to digital base reference maps will benefit and expedite review by the WBD–NTC.

The NGMC and the USGS will make the data and metadata available electronically to the member agencies of the SSWD for review and decisions about acceptance, when necessary. The process for the submittal, review, and acceptance is shown in figure 21. The data will be reviewed for adherence to the current "Federal Standards and Procedures for the National Watershed Boundary Dataset (WBD)" interagency guidelines, particularly the following:

- mapping and delineation accuracy and consistency;

- coordination of boundaries and codes that match across State or Subregion boundaries;

- hydrologic unit size, nesting, coding protocol (including format), and completion of required fields in the attribute table;

- completion of required metadata.

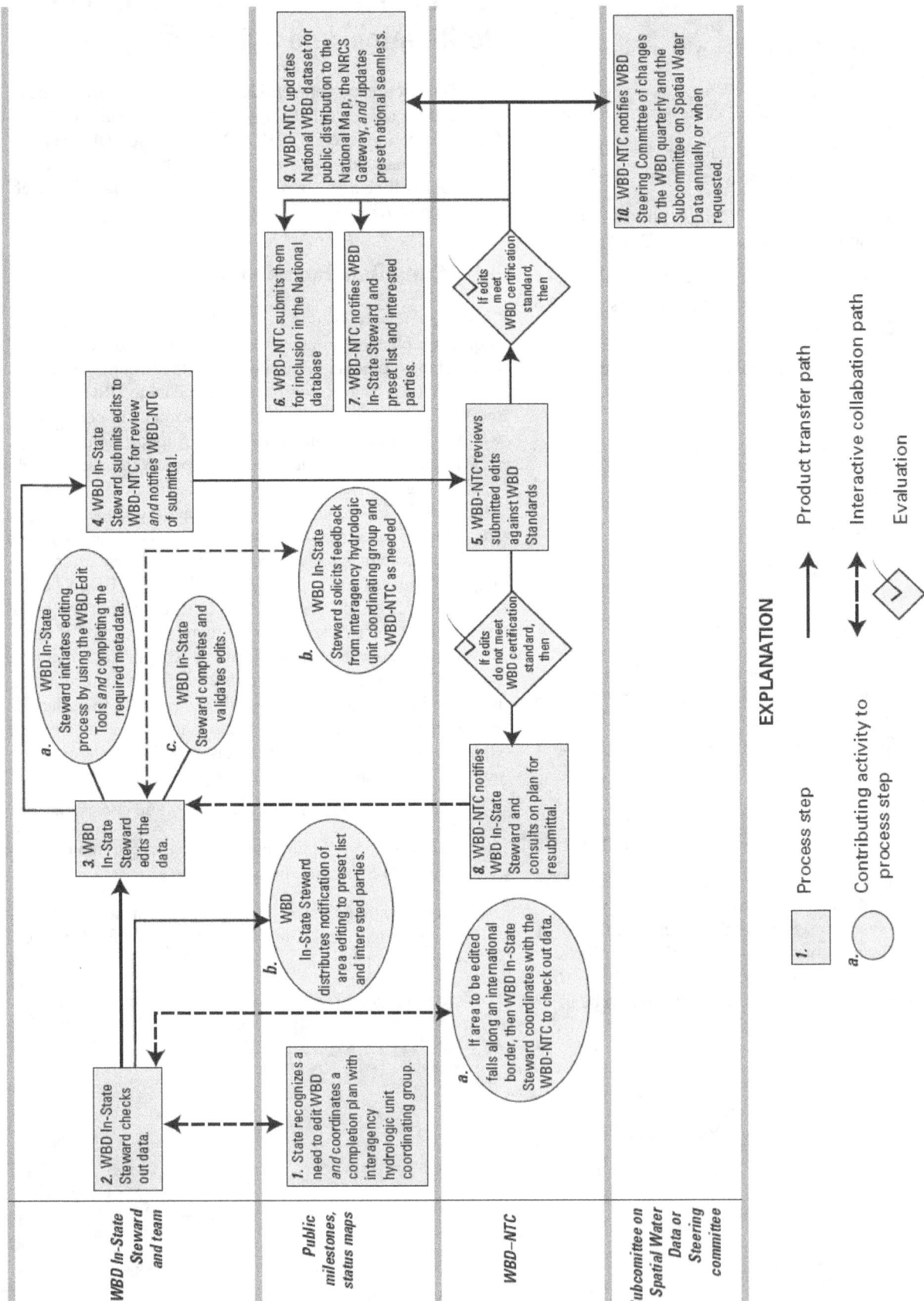

Figure 21. The Watershed Boundary Dataset revision process and review for national certification compliance and roles are depicted in the flowchart.

The WBD–NTC will refer review comments from the interagency team to the office that submitted the data. After discrepancies are resolved, the WBD–NTC will notify the WBD In-State Steward, specifying that the hydrologic unit data are approved and meet the national guidelines.

After the initial completion of WBD, some States were given a Provisional Certification status, which usually indicated that work such as providing metadata or matching boundaries across State lines remained to be completed. Provisional Certification was not given when obvious corrections needed to be made. If the data complied with the "Federal Standard for Delineation of Hydrologic Unit Boundaries" (Natural Resources Conservation Service, 2004) and were approved, the NGMC and the USGS sent a letter to the originating office stating that the WBD met the national standards, was fully certified, and would be added to the national database.

8.2 Data Checkout

WBD data can be checked out for editing from the NDH/WBD Stewardship Web site at *http://usgs-mrs.cr.usgs.gov/stewardbeta/index.html*. Every WBD In-State Steward will be required to obtain a valid username and password prior to login and checkout of data for editing. Contact the WBD-NTC for assistance in obtaining a login.

Once logged in, the WBD In-State Steward selects by 8-digit hydrologic unit code (Subbasin) the area(s) to be checked out and edited. Information related to the nature of the editing is required in the checkout form. The size of checkout should be limited to an area that can be successfully completed within a 90-day period.

When the checkout process is completed, the WBD In-State Steward will be notified by email that the job is ready for editing. Each job receives a unique job number, which is used to *"get"* the job from USGS servers prior to editing the data with the WBD Edit Tools.

8.2.1 Stewardship Web Site

An additional new site for WBD stewardship is the WBD stewardship Web site hosted by USGS *http://nhdpreview.er.usgs.gov/wbd.html*. This new site provides links for downloading WBD and NHD data, information about national and State WBD contacts, historical information, FGDC metadata, and other pertinent information.

8.3 Data Editing

The WBD In-State Steward begins the editing process by using the WBD Edit Tools. Instruction for using the WBD Edit Tools is provided on the WBD stewardship website at *http://nhdpreview.er.usgs.gov/wbd.html*. Users will also use the tools for populating related metadata, validating edits, and submitting data for national review by the WBD–NTC.

8.4 Data Verification

This section sets forth the rules that WBD In-State Stewards need to follow for handling data check-out and check-in through the USGS stewardship process.

8.4.1 File Names

All dataset checkouts are named according to USGS job process identification, and the dataset name should never be changed. Copying, projecting, or renaming checked out datasets will prevent them from being submitted successfully.

8.4.2 Map Projection and Horizontal Datum

Data submitted for certification must be in geographic coordinates stored in decimal degrees. The horizontal datum is the North American Datum of 1983 (NAD 83). All projection information must be included in the metadata file. Projecting a WBD dataset that has been checked out will prevent it from being submitted successfully. If areal calculations are required, a projected copy for personal reference may be needed.

8.5 Data Submittal

Data are submitted through the WBD Edit Tool by the WBD In-State Steward. The tools perform the appropriate compression and delivery of the data to USGS.

9. Metadata

According to an Executive Order 12906 signed by President William J. Clinton on April 11, 1994, all Federal agencies developing geospatial data are required to document newly created data by completing metadata. The most recent FGDC content standards for metadata must be followed. A sample of a completed FGDC-compliant metadata template for hydrologic units is available at *http://www.nrcs.usda.gov/wps/portal/nrcs/detail/national/water/watersheds/dataset/?&cid=nrcs143_021631*.

Beginning in the spring of 2012, all required metadata entry is collected at the beginning of an editing session within the WBD Edit Tools. Examples of metadata templates are available through the WBD Edit Tools.

10.   Selected References

Dana, P.H., 1995a, Coordinate systems overview: Boulder, Colo., University of Colorado at Boulder, Department of Geology, accessed October 23, 2007, at *http://www.colorado.edu/geography/gcraft/notes/coordsys/coordsys.html*.

Dana, P.H., 1995b, Geodetic datum overview: Boulder, Colo., University of Colorado at Boulder, Department of Geology, accessed October 23, 2007, at *http://www.colorado.edu/geography/gcraft/notes/datum/datum_f.html*.

Environmental Systems Research Institute, 2007, Data in the DBMS—Learn how data converts from one type to another: Redlands, Calif., accessed August 3, 2007, at *http://webhelp.esri.com/arcgisdesktop/9.2/index.cfm?TopicName=Data_types_in_the_DBMS*.

Federal Geographic Data Committee, 1998a, Content standard for digital geospatial metadata (revised June 1998): Washington, D.C., FGDC–STD–001–1998, 78 p., accessed February 14, 2007, at *http://www.fgdc.gov/standards/projects/FGDC-standards-projects/metadata/base-metadata/v2_0698.pdf*.

Federal Geographic Data Committee, 1998b, Geospatial positioning accuracy standards, pt. 3—National standard for spatial data accuracy: Washington, D.C., FGDC–STD–007.3–1998, 25 p., accessed February 14, 2007, at *http://www.fgdc.gov/standards/projects/FGDC-standards-projects/accuracy/part3/chapter3*.

Lamke, R.D., Brabets, T.P., and McIntire, J.A., 1994, Alaska hydrologic units (revised 1996): U.S. Geological Survey, scale 1:250,000, accessed February 14, 2007, at *http://agdc.usgs.gov/data/projects/anwr/datahtml/akhuc.html*.

National Oceanic and Atmospheric Administration, National Ocean Service, Office of Coast Survey, n.d., U.S. Maritime Zones/Boundaries: accessed December 22, 2008, at *http://www.nauticalcharts.noaa.gov/csdl/mbound.htm*.

Natural Resources Conservation Service, 2007, Watershed Boundary Dataset (WBD) 1: accessed October 23, 2007, at *http://www.ncgc.nrcs.usda.gov/products/datasets/watershed/*.

Seaber, P.R., Kapinos, F.P., and Knapp, G.L., 1987, Hydrologic unit maps: U.S. Geological Survey Water-Supply Paper 2294, 63 p., accessed February 14, 2007, at *http://pubs.usgs.gov/wsp/wsp2294/*.

Snyder, J.P., 1987, Map projections, a working manual: U.S. Geological Survey Professional Paper 1395, 383 p., available online at *http://pubs.er.usgs.gov/usgspubs/pp/pp1395*.

U.S. Bureau of the Budget, 1947, United States national map accuracy standards: Washington, D.C., accessed February 29, 2012, at *http://nationalmap.gov/standards/pdf/NMAS647.PDF*.

U.S. Geological Survey [n.d.], United States Geographic Names Information System (GNIS): accessed, at *http://geonames.usgs.gov/pls/gnispublic/f?p=gnispq:8*

U.S. Geological Survey [n.d.], United States Geographic Names Information System (GNIS) feature class definitions: accessed February 14, 2007, at *http://geonames.usgs.gov/domestic/feature_class.htm*.

U.S. Geological Survey, 1994, 1:250,000-scale hydrologic units of the United States: accessed February 14, 2007, at *http://water.usgs.gov/GIS/metadata/usgswrd/XML/huc250k.xml*.

U.S. Geological Survey, 1999a, Map accuracy standards: U.S. Geological Survey Fact Sheet FS–171–99, accessed October 23, 2007, at *http://erg.usgs.gov/isb/pubs/factsheets/fs17199.html*.

U.S. Geological Survey, 1999b, Standards for National Hydrography Dataset: accessed February 29, 2012, at *http://nationalmap.gov/standards/nhdstds.html*.

U.S. Geological Survey, 2005, Elevation Derivatives for National Applications (EDNA): accessed October 23, 2007, at *http://edna.usgs.gov/*.

U.S. Geological Survey, 2006a, Digital Elevation Models (DEMs): accessed March 13, 2007, at *http://edc.usgs.gov/products/elevation/dem.html*.

U.S. Geological Survey, 2006b, Digital Raster Graphics (DRGs): accessed March 13, 2007, at *http://topomaps.usgs.gov/drg/*.

U.S. Geological Survey, 2006c, National Elevation Dataset (NED): accessed March 13, 2007, at *http://ned.usgs.gov/*.

U.S. Geological Survey, 2006d, National Hydrography Dataset (NHD): accessed March 13, 2007, at *http://nhd.usgs.gov/*.

U.S. Geological Survey, 2007, The National Map—Orthoimagery layer: accessed October 23, 2007, at *http://erg.usgs.gov/isb/pubs/factsheets/fs20073008/index.html*.

U.S. Geological Survey, 2009, National Hydrography Database stewardship handbook: U.S. Geological Survey: accessed May 27, 2009, at *http://webhosts.cr.usgs.gov/steward/index.html*.

11. Appendixes

11.1 Hydrologic Unit Names, Historical Names, Average Sizes, and Approximate Number of Hydrologic Units, Per Nested Level, in the United States

The historical names of the general classes of hydrologic units are being replaced in the Geographic Name Information System (GNIS) with numerical names that refer to the number of digits that indicate the place of a hydrologic unit in the nested hierarchy of drainage areas organized by size and position. This publication uses the same numerical names for hydrologic units that will be used in the GNIS.

Hydrologic unit name	Historical name	Average size (square miles)	Approximate number of hydrologic units
2 digit	Region	177,560	21 (actual)
4 digit	Subregion	16,800	222
6 digit	Basin	10,596	370
8 digit	Subbasin	700	2,270
10 digit	Watershed	227 (40,000–250,000 acres)	20,000
12 digit	Subwatershed	40 (10,000–40,000 acres)	100,000
14 digit	(None)	Open	Open
16 digit	(None)	Open	Open

11.2 Definitions

The hydrologic (Office of Water Research and Technology, 1980) and compliance terms in the following lists are defined with reference to the WBD process. The definitions may not be the only valid ones for these terms. Boldfaced terms within definitions are defined elsewhere in the lists.

11.2.1 Hydrologic Definitions

14-digit hydrologic unit   A subdivision of a 12-digit **hydrologic unit** (Subwatershed).

16-digit hydrologic unit   A subdivision of a 14-digit **hydrologic unit**, and the smallest of the hydrologic unit hierarchy.

Basin   A subdivision of **Subregion**. A Basin is the third-level, 6-digit unit of the **hydrologic unit** hierarchy. Basins were formerly named "accounting units" in USGS terminology.

Composite hydrologic unit   A topographically defined area where all the surface drainage converges to a single point, usually along the main stem of a stream between outlets of **standard hydrologic units**. This includes areas or small triangular wedges (**remnant areas**) between adjacent drainage areas that remain after standard hydrologic units are delineated.

Contiguous boundary   A **hydrologic unit** boundary shared in whole or in part by an adjacent, different hydrologic unit.

Estuary   The region of interaction between streams and nearshore ocean waters, where tidal action and streamflow mix freshwater and saltwater.

Frontal hydrologic unit   A land and water area where surface flow originates entirely within the **hydrologic unit** and drains to multiple points along a large waterbody, such as the ocean or large lake.

Head of land   A projection of land extending into a waterbody that interrupts the coastal trend of that waterbody. The point where the ridgeline meets the waters' edge is connected to a similar point on the opposite bank of the channel to form a boundary across the mouth of an outlet.

Hydrography   The scientific description, study, and analysis of the physical conditions, boundaries, measurement of flow, investigation and control of flow, and related characteristics of surface water such as streams, lakes, and oceans.

Hydrologic unit (HU)   An identified area of surface drainage within the United States system for cataloging drainage areas, which was developed in the mid-1970s under the sponsorship of the Water Resources Council and includes drainage-basin boundaries, codes, and names. The drainage areas are delineated to nest in a multilevel, hierarchical arrangement. The hydrologic unit hierarchical system has four levels and is the theoretical basis for further subdivisions that form the **Watershed Boundary Dataset** fifth and sixth levels. A hydrologic unit can accept surface water directly from upstream drainage areas and indirectly from associated surface areas, such as **remnant areas, noncontributing areas**, and diversions, to form a drainage area with single or multiple outlet points.

Hydrologic unit code (HUC)   The numerical identifier of a specific **hydrologic unit** or drainage area consisting of a two-digit sequence for each specific level within the delineation hierarchy.

Hydrologic unit name   A standardized name assigned to a **hydrologic unit** used to identify the geographic location of the area. Hydrologic units are typically named after significant or prominent water features in an area; however, in some instances, they may also be named after other features.

Hydrology   The science dealing with the properties, distribution, and circulation of water on the surface of the land, in the soil and underlying rocks, and in the atmosphere.

Karst areas   Areas of carbonate-rock formations (limestone and dolomite) characterized by sinks, underground streams, and caverns.

Nearshore boundary   An offshore closure line for **hydrologic units**, determined by State subject matter experts, based on a depth or a distance from the mean high water designation or legally defined lake shoreline. (The outermost closure line is the NOAA Three Nautical Mile Line required for open-ocean hydrologic units.)

Noncontributing area   A naturally formed area that does not contribute to the downstream accumulation of streamflow under normal flow conditions.

Offshore boundary   Any boundary in water, including near shore, head of land, the NOAA Three Nautical Mile Line, and adjustments or generalizations where applying a buffer or distance guideline is problematic.

Open-water hydrologic unit   An area delineated within an ocean. Land is not a major portion of the **hydrologic unit,** but land may be included, as in the case of islands.

Region   A Region is the first-level, 2-digit unit and is the largest in the **hydrologic unit** hierarchy.

Remnant area   A topographically defined area that is residual after delineation of **standard hydrologic units**. Remnant areas include small triangular wedges between adjacent drainage areas. These areas are commonly incorporated into **composite hydrologic units**.

Shoal   A natural accumulation of sand, gravel, or other material forming a shallow underwater or exposed embankment.

Sound   (*A*) A relatively narrow sea or stretch of water between two close landmasses that connects two larger bodies of water. (*B*) A deeper part of a moving body of water (as bays, estuaries, or straits) through which the main current flows or that affords the best passage through an area otherwise too shallow to navigate.

Standard hydrologic unit   An area with drainage flowing to a single outlet point. Examples include **composite** and **remnant hydrologic units**.

Subbasin   A subdivision of a **Basin**. A Subbasin is the fourth-level, 8-digit unit of the **hydrologic unit** hierarchy. Subbasins were formerly named "cataloging unit" in USGS terminology. The average size is about 450,000 acres.

Subregion   A subdivision of a **Region**. A Subregion is the second-level, 4-digit unit of the **hydrologic unit** hierarchy. The hydrologic unit category name is retained for the **Federal Standards and Procedures for the National Watershed Boundary Dataset (WBD).**

Subwatershed   A subdivision of a **Watershed**. A Subwatershed is the sixth-level, 12-digit unit of the **hydrologic unit** hierarchy. Subwatersheds generally range in size from 10,000 to 40,000 acres.

Toe of the shore face   A demarcation depth to which seasonal storms, prevailing winds, and resultant waves and currents move shallow sediments to and from the shore. From this geomorphic feature toward the shore, water depth decreases rapidly for a short distance and then slowly for the remaining distance.

True hydrologic unit *See* **standard hydrologic unit**.

Water hydrologic unit   A body of water that can receive flow from adjacent **frontal hydrologic units, composite hydrologic units**, and **standard hydrologic units**. Examples include inland lakes and nearshore ocean waters. Generally, these units are not delineated to the next lower **hydrologic unit** level.

Watershed   (*A*) In the hierarchy of hydrologic units, a 10-digit hydrologic unit (fifth level) also is known as a "Watershed," and it is a subdivision of an 8-digit (fourth level) unit, also known as a "**Subbasin**." These 10-digit hydrologic units range in size from 40,000 to 250,000 acres. (*B*) The hydrologic term "watershed" refers to the divide that separates one drainage basin from another or to a combination of hydrologic units of any size.

11.2.2   Geospatial Data Definitions and Standards

11.2.2.1   Definitions

Attribute   A defined characteristic of a geographic feature or entity.

Base classes   The points, lines, attributes and related tables that are updated during the editing process (as defined by the USGS National Geospatial Program Office).

Contour line   A line (as on a map) connecting the points on a land surface that have the same altitude.

Coordinates   Pairs of numbers expressing horizontal distances along orthogonal axes; alternatively, triplets of numbers measuring horizontal and vertical distances.

Coordinate system   A system in which points on the Earth's surface are located with reference to a pair of intersecting lines or grid. For more information, see Dana (1995a) and Snyder (1987).

Compilation   The act of composing new or revised materials from existing documents or sources.

Crosswalk (Crosswalking)   The use of a table to show the relationship between elements in other tables.

Dataset   A collection of related data.

Datum   A reference surface for a geodetic survey. Refers to a direction, level, or position from which angles, heights, depths, and distances are normally measured. Datum, as applied to a horizontal geodetic survey, is a reference based on the shape of the Earth. For more information, see Dana (1995b).

Delineation   The act of indicating or representing by drawn lines.

Derived classes   The attributed polygons, organized by hydrologic unit level, starting with the 2-digit hydrologic units and ending with the 16-digit hydrologic units (as defined by the U.S. Geological Survey National Geospatial Program Office). The derived classes also include the attributed polygons of the NWIS drainage areas.

Edge matching   A digital editing procedure equivalent to joining adjacent features at hardcopy map edges; used to ensure that features crossing political boundaries or adjoining maps connect.

Geographic information system (GIS)   A computer system designed to collect, manage, manipulate, analyze, and display spatially referenced data and associated attributes.

Geospatial data   Information that identifies the geographic location (coordinate system) and characteristics (attributes) of natural or constructed features and boundaries on the Earth. The numerical scale associated with geospatial data refers to the spatial accuracy, the smallest scale of delineation, and the scale of other data that are spatially compatible.

Hypsography   The study of the distribution of elevations on the surface of the Earth with reference to a datum, traditionally sea level.

Metadata   The description and documentation of the content, quality, condition, and other characteristics of geospatial data.

Polygon   A sequence of alternating line segments and angled vertices that form a closed two-dimensional loop, thus defining the boundary of an area.

Resolution   The minimum difference between two independently measured or computed values that can be distinguished by the measurement or analytical method being considered or used.

Topology   The spatial relations between geometric entities, including adjacency, containment, and proximity.

Vector data   A **coordinate**-based data structure used to represent positional data in spatial units of line, point, and **polygon**.

Vector digitizing   The act of tracing a line with a device to capture and store the locations of geographic features by converting their map positions to a series of x-y coordinates.

Watershed Boundary Dataset (WBD)   A national geospatial database of drainage areas consisting of the 2- through12-digit hydrologic units, with optional 14- and 16-digit hydrologic units. The WBD includes the required **attribute** and **metadata** information. For more information, see Natural Resources Conservation Service (2007).

11.2.2.2   Standards

Federal Geographic Data Committee (FGDC)   The FGDC develops geospatial data standards for implementing the nationwide data publishing effort known as the National Spatial Data Infrastructure (NSDI). For the most recent information, see http://www.fgdc.gov/standards.

National Map Accuracy Standards (NMAS)   Specifications formerly governing the accuracy of topographic, base,

orthophoto, and other maps produced by Federal agencies. For more information, see U.S. Geological Survey (1999a).

National Standards for Spatial Data Accuracy (NSSDA)   Specifications superseding the NMAS as the standard for governing the accuracy of topographic, base, orthophoto, and other maps produced by Federal agencies. The **WBD** is **delineated** and georeferenced to USGS 1:24,000-, 1:25,000-, or 1:63,360-scale topographic quadrangle maps that meet the most current NSSDA standard. For more information, see Federal Geographic Data Committee (1998a,b).

11.2.3   Compliance Definitions

Certification   Formal acknowledgment that **hydrologic units** have been reviewed and meet the criteria as stated in the "Federal Standards and Procedures for the National Watershed Boundary Dataset (WBD)," 2011, as agreed to by member agencies of the SSWD.

Provisional Certification   Acknowledgment that **hydrologic units** for large regional areas or whole States have been reviewed and meet the criteria as stated in "Federal Standards for Delineation of Hydrologic Unit Boundaries; Version 2.0 October 1, 2004," as agreed to by member agencies of the SSWD. Hydrologic unit boundaries and attributes are **edge matched** to the fullest possible extent at that time. States are encouraged to work with neighboring States to resolve boundary issues and resubmit the data to attain final **certification**.

Prior to April 2006, Provisional Certification was known as Certification or Verification.

Full Certification   **Hydrologic units** have been reviewed, and meet the National requirement as stated in "Federal Standards for Delineation of Hydrologic Unit Boundaries; Version 2.0, October 1, 2004," agreed to by member agencies of the SSWD.

11.3   Abbreviations and Acronyms

ACWI	Advisory Committee on Water Information
BLM	Bureau of Land Management
DEM	Digital Elevation Model
DOQ	Digital Orthophoto Quadrangle
DOQQ	Digital Orthophoto Quarter Quadrangle
DRG	Digital Raster Graphic
EDNA	Elevation Derivatives for National Applications
Esri	Environmental Systems Research Institute
FGDC	Federal Geographic Data Committee
FTP	File Transfer Protocol
GIS	Geographic Information System
GNIS	Geographic Names Information System
HUC	Hydrologic unit code
IHUCG	Interagency Hydrologic Unit Coordinating Group
IFSAR or IfSAR	Interferometric synthetic aperture radar
INEGI	Instituto Nacional de Estadística y Geografía
LIDAR or LiDAR	Light detecting and ranging
NED	National Elevation Dataset
NGMC	National Geospatial Management Center (formerly National Geospatial Cartographic Center (NCGC))
NGTOC	National Geospatial Technical Operations Center
NHD	National Hydrography Dataset
NMAS	National Map Accuracy Standards
NOAA	National Oceanic and Atmospheric Administration
NRCS	Natural Resources Conservation Service (U.S. Department of Agriculture)
NSSDA	National Standards for Spatial Data Accuracy
QA/QC	Quality assurance/quality control
SSWD	Subcommittee on Spatial Water Data
STDS	Spatial Data Transfer Standard
USDA	U.S. Department of Agriculture
USDA–FS	U.S. Department of Agriculture, Forest Service
EPA	U.S. Environmental Protection Agency
USGS	U.S. Geological Survey
WBD	Watershed Boundary Dataset
WBD–NTC	Watershed Boundary Dataset National Technical Coordinators

11.4   Data Type Crosswalk for Data Storage of Esri Products

SQL standard	Arc GIS	Esri INFO (coverage)	Esri ArcView (shape file)	Esri ArcSDE personal (access)	Esri ArcSDE file based	Esri ArcSDE enterprise			
						Oracle	MS SQL (server)	DB2	Informix
EXACT NUMERIC	SHORT INTEGER	I	Number, Boolean	Integer	SHORT INTEGER	NUMBER (4)	SMALLINT (2)	SMALLINT (2)	SMALLINT (2)
EXACT NUMERIC	LONG INTEGER	B,I	Number	Long integer	LONG INTEGER	NUMBER (38)	INT (4)	INTEGER (4)	INTEGER (4)
APPROXIMATE NUMERIC	FLOAT	F,N	Number	Single	FLOAT	NUMBER (38,8)	REAL	DECIMAL (31,8)	DECIMAL (32)
APPROXIMATE NUMERIC	DOUBLE	F,I,N	Number	Double	DOUBLE	NUMBER (38,8)	DOUBLE	DECIMAL (31,8)	DECIMAL (32)
CHARACTER STRING	TEXT	C	String	Text	TEXT	VARCHAR	VARCHAR (N)	VARCHAR	VARCHAR (N)
DATE TIME	DATE	D	Date	Date/time	DATE	DATE	DATETIME	TIMESTAMP	DATETIME
BIT STRING	BLOB	—	—	OLE object	BLOB	BLOB	IMAGE	BLOB	BLOB
EXACT NUMERIC	GUID	—	—	Number	GUID	CHAR (38)	UNIQUE IDENTIFIER (16)	CHARACTER (38)	CHAR (8)

11.5   Previously Used Metadata Identification Attribute for Tracking Metadata

Attribute	Field name	Field type	Field width	Domain	Is value required ?	Example
Metadata identification	MetaID	Character	4	AL, AK, AR, AZ, CA, CO, CT, DE, FL, GA, HI, IA, ID, IL, IN, KS, KY, LA, MA, MD, ME, MI, MN, MO, MS, MT, NC, ND, NE, NH, NJ, NM, NV, NY, OH, OK, OR, PA, RI, SC, SD, TN, TX, UT, VA, VT, WA, WI, WV, WY, PR, VI, AS, FM, GU, MH, MP, PW, UM;01-99	Yes	OK01

11.6 Historical Sample Letter for Certification or Data Update Submittal

To help the review and certification process, consider including these categories of information about the history and development of the dataset in the transmittal letter. Contact the WBD–NTC for questions or assistance.

Date:

Subject: Submittal of the Watershed Boundary Dataset for

To: Director
        USDA Natural Resources Conservation Service
        National Geospatial Management Center (NGMC)
        501 West Felix Street, Building 23 Fort Worth, TX 76115

From: [Name]
        [Title]
        [Agency and address]

I am pleased to inform you that [State] has completed its Watershed Boundary Dataset (WBD) in accordance with the [standards document used]. This dataset, along with its metadata, has been posted at the following FTP site for retrieval: Describe the timeframe, major cooperators and agencies involved in development, and formal agreements in place. For example,

"The dataset was created by the coordinated effort between multiple agencies. NRCS provided the oversight and did a major portion of the delineations and attribution in [specify] part of the State. Over [timeframe] [other agencies and locations] in [State] as well as [WBD–NTC agent] have, under an agreement between [agencies] and [State], completed the remainder of the State's boundaries and attributes, and appended the dataset together."

Describe general production methods and reference or source materials used. For example,

"This dataset was produced digitally on screen in ArcView or ArcMap, referencing the Digital Raster Graphics for the method of delineation. As the dataset evolved, and especially in the northeastern portion of the State, the National Hydrography Dataset, along with the Digital Ortho Quarter Quads (DOQQs), provided very useful as a reference for determining breaks in the flood plains. DOQQs were cited frequently along the [specify major features] as well. Every effort has been made to create new boundaries and revise the historical linework to meet the National Standards for Spatial Data Accuracy (NSSDA) for 1:24,000 meter scale."

Describe edge-matching activities and outcomes. Include desired scenario involving States without current WBD. For example,

"Edge-matching with [list surrounding States] has been completed. [List States without existing WBD delineation]. [Explain desired scenario involving States without current delineation]. It was agreed upon with adjacent States that [your State] would delineate to the [special areas] and the bordering States would include the remainder of the [State] Subbasin boundaries within those States WBDs." Itemize other unique areas and circumstances.

Final QA/QC was performed by [WBD–NTC agent] under this latest agreement, as well as [State interagency hydrologic unit group or lead agency].

If there should be any technical questions during the review process, or edits to be made, please feel free to contact me (information listed below). Or you may contact [WBD–NTC agent] by [phone] or [email]

Sincerely,

[Name]

[Agency and address]

cc: Please fill in list of major people to notify

www.ingramcontent.com/pod-product-compliance
Lightning Source LLC
Chambersburg PA
CBHW081603170526
45166CB00009B/2800